# Editorial

# Challenging Nuclearism

A cruel year finally closed. The COVID-19 Pandemic has coincided with the deaths of many friends and supporters of the Russell Foundation. In April 2020, we lost Ken Fleet, who had been the Foundation's tireless Secretary for 50 years. The year also claimed Robert Fisk and John Le Carré, contributors to *The Spokesman*. In this issue, we remember two other contributors who died in 2020—Mike Cooley and Stephen Cohen, who made their mark in different fields.

I first heard Mike Cooley speak at an Institute for Workers' Control conference in the early 1980s. Mrs Thatcher was Prime Minister and the prospects were bleak. Mike spoke about tacit knowledge — how workers know what to do. He probed hand-eye co-ordination, going on to celebrate the skill entailed in safely crossing a busy road. When, in later years, I reminded him of this example he had given, he was somewhat surprised, though it did stir a memory. For he would occasionally phone the Foundation to keep in contact, particularly after we republished *Architect or Bee?* in 2016. Mike then sent the manuscript of *Delinquent Genius*, and asked for an opinion. He had deliberated about it for some decades. My immediate response was that these insights, personal and scientific, should be shared more widely. So it was that the book came to be published in 2018, with a generous introduction by Michael D Higgins, President of Ireland and Mike's old school friend from Tuam in County Mayo. We reprint a chapter from that book, together with John Palmer's appreciation of Mikes life, which is the Foreword to *The Search for Alternatives*, the third title in the Cooley trilogy published by Spokesman. Mike's archive now resides at Waterford Institute of Technology, where it attracts international attention. Meanwhile, the Lucas Plan discussion network links Mike's insights to ongoing work for a Green New Deal and related questions. Closer to home, Shirley, Mike's wife, and Graham, his son, keep the flame burning.

Stephen Cohen was another longtime friend of the Russell Foundation. He collaborated with Ken Coates in the extended campaign, ultimately successful, for the rehabilitation of Nikolai Bukharin in the Soviet Union. Bukharin's widow, Anna Larina, and son, Yuri, had exhausted every official avenue in the Soviet Union, so that a wider international campaign was necessary. John Daniels sketches Cohen's broader achievements in explaining modern Russia to the wider world.

\* \* \*

The Treaty on the Prohibition of Nuclear Weapons came into force in January 2021. It is a notable achievement, the result of long decades of campaigning at the United Nations and in wider civil society. We publish a Dossier which includes the text of the Treaty itself, together with critical commentaries including one by Richard Falk, which gives us our title for this issue. The Dossier was compiled by Tom Unterrainer, who has kindly agreed to join me as co-editor of *The Spokesman* journal. Tom piloted into print our little book entitled *Why Trident?* From this, we have selected excerpts by Vice Admiral Blackham and Commander Forsyth.

\* \* \*

The transition period has ended and people in the UK endure the consequences of 'Hard Brexit', embodied in the minimal trade agreement with the European Union signed by Prime Minister Boris Johnson. Travel and trade have become more difficult, and some 60 million UK nationals have apparently lost their status as EU citizens, although this is being challenged at the European Court of Justice. Meanwhile, millions of EU nationals resident in the UK, together with millions more UK nationals resident in continental Europe, have endured life in Limbo since the Brexit referendum of 2016. We reprint first-hand accounts of what this really means from *In Limbo: Brexit Testimonies from EU Citizens in the UK*, compiled by Elena Remigi and colleagues. We're pleased to publish this landmark work under the Foundation's imprint, together with *In Limbo Too: Brexit Testimonies from UK Citizens in the EU*. We are also grateful to PM Press for permission to reprint a chapter from their landmark title *Asylum for Sale: Profit and Protest in the Migration Industry*.

\* \* \*

Banksy visited Nottingham, once known for its cycle industry, and brightened a street corner where children play. Whilst browsing his website (banksy.co.uk), we found a cover for *Challenging Nuclearism* which made us smile. We hope pest control office approves.

*Tony Simpson*

# The Spokesman
## Challenging Nuclearism
Edited by Tony Simpson & Tom Unterrainer

Published by Spokesman for the
Bertrand Russell Peace Foundation
Ken Coates: Editor 1970 to 2010

Spokesman 147     2021

**Subscriptions**
Institutions £40.00 (ex UK)
£33.00 (UK)
Individuals £20.00 (UK)
£25.00 (ex UK)

A CIP catalogue record for this book is available from the British Library

### CONTENTS

| | | |
|---|---|---|
| Editorial | 3 | |
| Nothing sacrosanct | 7 | *Vice Admiral Sir Jeremy Blackham* |
| UK Nuclear Weapon Policy | 9 | *Cdr Robert Forsyth* |
| **TPNW Dossier** | | |
| Through the doorway | 21 | *Setsuko Thurlow* |
| Turning Point | 24 | *Daryl G. Kimball* |
| Changing Europe's Calculations | 30 | *Beatrice Fihn and Daniel Högsta* |
| Treaty on the Prohibition of Nuclear Weapons | 34 | |
| Disarmament | 49 | *Bertrand Russell* |
| Challenging Nuclearism | 53 | *Richard Falk* |
| Rely on science | 61 | *José Bustani* |
| In Limbo | 65 | *Elena Remigi et al* |
| Human Skill | 70 | *John Palmer* |
| The Shout of Joy | 73 | *Mike Cooley* |
| Renewable Energy | 82 | *Dexter Whitfield* |
| Asylum For Sale | 86 | *Alva White, Uyi & Madi* |
| Stephen F. Cohen | 98 | *John Daniels* |
| The Claims of Women | 103 | *Kate Amberley* |
| Reviews | 120 | *Stan Newens* |
| | | *Helen Jackson* |
| | | *Stephen Winfield* |
| | | *Louise Regan* |
| | | *Anthony Lane* |
| | | *Andrew Bone* |

Cover: Banksy

ISSN 1367 7748     ISBN 978 0 85124 8950

**Published by**
The Bertrand Russell Peace Foundation Ltd,
5 Churchill Park,
Nottingham, NG4 2HF
England
Tel. 0115 9708318
email: editor@russfound.org
www.spokesmanbooks.com
www.russfound.org

**Editorial Board**
John Daniels
Kate Fleet
Stuart Holland
Henry McCubbin
Abi Rhodes
Regan Scott

Mixed Sources
Product group from well-managed forests and other controlled sources
Cert no. SGS-COC-006541
www.fsc.org
© 1996 Forest Stewardship Council

In addition to social changes designed to bring security there is, however, another and more direct means of diminishing fear, namely by a regimen designed to increase courage. Owing to the importance of courage in battle, men early discovered means of increasing it by education and diet ... But military courage was to be the prerogative of the ruling caste: Spartans were to have more than helots, British officers than Indian privates, men than women, and so on. For centuries it was supposed to be the privilege of the aristocracy. Every increase of courage in the ruling caste was to increase the burdens on the oppressed, and therefore to increase the grounds for fear in the oppressors, and therefore to leave the causes of cruelty undiminished. Courage must be democratised before it can make men humane.

**Bertrand Russell**, *What I Believe* (1925)

Image by Rufus Segar from *Russell and the Anarchists, Anarchy* No. 109, March 1970

# WHY TRIDENT?

## Commander Robert Forsyth RN (Ret'd)

Foreword by Vice Admiral Sir Jeremy Blackham KCB MA
Introduction by Professor Nick Grief BA PhD Barrister

# Nothing sacrosanct

*Vice Admiral Sir Jeremy Blackham KCB MA*

*Vice Admiral Blackham served as Deputy Commander-in-Chief, Fleet and, in retirement, was Editor of* The Naval Review. *This is his Foreword to Commander Robert Forsyth's book,* Why Trident? *(Spokesman Books, 2020)*

The British submarine-borne nuclear deterrent was first deployed in 1969, at the height of the Cold War, and has been an unchallengeable cornerstone of British defence and security ever since. But in that time much has changed. Many more countries now possess, or have the means to possess, nuclear weapons; international relationships are different and more complex; critical threats emanate from non-state actors. British international influence has declined, not least because of her reduced conventional military power and the growth of new forms of warfare, perhaps less amenable to nuclear dissuasion. However, the conventional forces that Britain has chosen to retain are now sufficiently reduced to lower very significantly the nuclear threshold, the point at which a decision to use nuclear weapons or rather to seek an accommodation is reached. Some people, of whom I am one, believe that a lower nuclear threshold greatly undermines the credibility of our nuclear deterrent because an adversary with far greater conventional and nuclear capability may not believe we would use it, or may conclude that he can win a conventional war before he has exhausted his conventional options or before we have found ourselves at a level of threat which could justify unleashing mutual nuclear destruction. Deterrence would have failed before nuclear forces came into play. One might even argue that we have deterred ourselves rather than the enemy.

Deterrence is an intellectually challenging subject and one in which no nation can afford to act without careful consideration of the interests of its allies and even those of its potential rivals. In the interconnected world in which we live, and the many natural and technological challenges, as well as the more conventional military threats we face, it is impossible to over-emphasise the value of being able to deter at any level of warfare, to prevent us reaching the nuclear threshold. Against this background, the United Kingdom has declared its intention to carry out a fundamental defence and security review, with the aim of reshaping its defence and security posture to meet the new challenges and those of the next generation. This is crucially important in an increasingly dangerous world.

Such a review must be a 'clean sheet' review; nothing should be sacrosanct. It must produce a policy which is coherent 'end-to-end' and equipped with the best and most effective tools we are prepared to afford. It is therefore very timely to review the nuclear element of our deterrent posture, in the light of the moral, legal, economic, political, environmental and practical issues involved. We can no longer simply assume that it is in all circumstances essential, irrespective of its impact on our overall security. Few people have examined more rigorously the critical questions surrounding the nuclear element of this 'continuum of deterrence' than Rob Forsyth. I am very glad that his work in this field has been collected into this book. All those involved in this field should read and carefully reflect on what he has to say.

# UK Nuclear Weapon Policy

*Commander Robert Forsyth*

*Commander Forsyth RN (Ret'd) served in the Royal Navy from 1957 to 1981. He commanded conventional and nuclear powered submarines, was Executive Officer of a Polaris missile equipped submarine, and led the Submarine Commanding Officers' Qualifying Course. This article is from his recent book,* Why Trident?, *published by Spokesman.*

**Is *Trident* independent?**

The justification for an 'independent deterrent' is that the UK must be able to use it entirely alone without US approval. The Government makes the following three assurances:

- "decision making and use of the system remains entirely sovereign to the UK; only the Prime Minister can authorise the launch of nuclear weapons, which ensures that political control is maintained at all times."
- "the instruction to fire would be transmitted to the submarine using only UK codes and UK equipment; making the command and control procedures fully independent."
- "our procurement relationship with the US regarding the *Trident* missile does not compromise the operational independence of our nuclear deterrent."[1]

All three beg the question as to whether the US can stop the UK from firing. The reality is that UK independence exists only so long as the US permits it. The Trident Commission – an authoritative, independent, cross-party inquiry which examined UK nuclear weapons policy – in its July 2014 Concluding Report stated that if the US withdrew support, UK 'independence' "would have a life expectancy measured in months".[2]

Dr Dan Plesch describes in considerable detail the extremely high level of UK dependence on the US, and the physical measures that the US could take to prevent a UK missile firing if it disapproved.[3] The missiles are maintained by, and leased from, a joint US-UK pool in Kings Bay, Georgia. The *Trident* replacement submarine missile tube module and its associated launch system is a joint project to be incorporated into the design of both the *Columbia* and *Dreadnought* class SSBNs. The onboard hardware and software systems to target the missiles are US supplied and maintained. Optimally they rely on US satellite-derived navigation and weather information for warhead guidance, albeit that less accurate fall-back systems can be used. Consequently the availability and use of the *Trident* weapon system is heavily reliant on US support and software skills. The warheads are notionally British, but US companies are deeply embedded in their design, and 70% of the company managing the Atomic Weapons Establishment (AWE) Aldermaston is US owned.[4] In sum, should the US wish to prevent the UK using *Trident*, it has the ability to do so.

Plesch points out that it is not inconceivable that the US, in the last resort, would consider military action to inhibit UK use. While this might seem incredible, the US was quite prepared to do so to stop the 1956 Anglo-French Suez campaign. General Sir Charles Keightley, UK Commander of Middle East Land Forces at the time, wrote afterwards: "It was the (military) action of the US which really defeated us in attaining our object." He complained that the actions of the US Sixth Fleet "endangered the whole of our relations with that country".[5]

In May 2019 there was a clear indication that the US is prepared to threaten reprisals on the UK if it does not comply with its wishes. The US Secretary of State, Mike Pompeo, warned that UK-US defence cooperation would be put at risk if the UK gave the Chinese company Huawei a role in operating the UK's 5G communication infrastructure.[6]

The Royal Navy (RN), Ministry of Defence (MoD) officials and Ministers all understand that maintaining the UK 'deterrent' as an effective weapon system is entirely dependent on US goodwill. As the former Prime Minister Tony Blair admitted in his autobiography: "[I]t is quite inconceivable that we would use our nuclear deterrent alone, without the US."[7] At a conference in June 2018, hosted by the National Museum of The Royal Navy, numerous RN and MoD speakers emphasised the dependence on the US for the effective operation of the UK *Trident* submarine force.

The illusion of an 'independent deterrent' is presented as fundamentally linked to UK permanent membership of the UN Security Council and thus a 'seat at the top table' as a major power. However, as one of the victors in World War II, the UK's membership was established before acquiring nuclear weapons; so this is irrelevant to its nuclear status. In support of one speaker's view at the 2019 Annual UK Project On Nuclear Issues (PONI) Conference that "UK possession of nuclear weapons has always been driven by the need for strong strategic links with the US", four recent occasions where the UK exactly shadowed the US position were pointed to. These were at conferences addressing the humanitarian impact of nuclear weapon use in Oslo (2013) and Vienna (2014), and the last two Nuclear Non-Proliferation Treaty (NPT) Review Conferences (2015, 2019).[8]

**The Cost**

In 2018 the total financial cost of replacing *Trident* was estimated at over £43Bn.[9] This makes the *Dreadnought* programme the second largest public capital procurement programme in the next decade, comparable only to the High Speed 2 railway line from London to Manchester and Leeds.

However, the full cost extends to the effect it has had on the operational capabilities of the Forces, and especially the RN. To accommodate this, the RN's conventional capabilities have been cut to the point where it would struggle to fulfil its historic core role of providing graduated conventional maritime deterrence. The current surface escort order of battle comprises six destroyers and 13 frigates – figures which match the six ships sunk in the Falklands War and 13 sufficiently damaged to put them out of action or severely limit their use. To put this in context, Rear Admiral Sir Sandy Woodward, the Operational Commander of all surface ships, land and air forces, stated: "During that time I lost nearly half of the destroyers and frigates I started with."[10] This was against a relatively limited enemy, engaging UK forces at long distance. Fortunately he had the numbers to absorb the high attrition rate. Similarly, on any given day only one, or possibly two, nuclear attack submarines are currently available – sometimes none – while the SSBN on Continuous At Sea Deterrence (CASD) deployment is a major liability requiring scarce ships and submarines to protect it as a very high value target. There is little or no provision for an attrition reserve today. Nelson famously said, "Were I to die at this moment, want of frigates would be found stamped on my heart."[11] Nothing has changed.

The financial and operational burdens of sustaining *Trident* are so great, and increasing, that they prejudice not just *Trident* renewal but the entire UK submarine-based nuclear weapons programme.[12] Some argue that this could be solved by moving the cost of renewing *Trident* back to the National Budget where it lay prior to 2010.[13] This would expose all the factors rehearsed here to the public, such that the political impact on the

NHS and other social budgets would not be acceptable. So instead the Government has been putting more pressure on the Navy to find savings elsewhere.

The negative consequences of acquiring *Polaris*, and subsequently replacing it with *Trident*, were foreseen by two First Sea Lords. Admiral of the Fleet Sir Caspar John, First Sea Lord in 1964, on learning of the *Polaris* Sales Agreement, warned of the "millstone of *Polaris* hung around our necks" and as "potential wreckers of the real Navy." Admiral of the Fleet Sir Henry Leach echoed his predecessor's warning by describing the *Trident* programme as "the cuckoo in the nest".

Vice Admiral Sir Jeremy Blackham, in his Foreword to Cdr Green's book *Security without Nuclear Deterrence*, correctly summed up the current situation: "But the cardinal point is that the nuclear deterrent is not and cannot be a substitute for conventional capabilities. The credibility of flexible response depends upon deferring any decision to use nuclear weapons until the very existence of the nation is at stake. This requirement means that the conventional forces must be of sufficient capability to deal with any lesser threat; and that one's potential enemy must believe this to be so." He further emphasised that "[i]f the conventional means at our disposal are weak, the point of transition to nuclear use may be lowered to levels at which the threat of nuclear obliteration is self-evidently wholly disproportionate ... At that point it is likely that deterrence through the threat of nuclear use becomes overtly incredible".[14]

## Continuous at Sea Deterrence (CASD)

The Government states that "invulnerability and security of capability are key components of the credibility of our deterrent and contribute to overall stability."[15] CASD is a hangover from the Cold War's perceived need to be able to respond immediately if subjected to a 'bolt from the blue' attack from the USSR. This is why the *Polaris* force was kept at 15 minutes' notice to fire. No such need has existed since 1994 when UK and Russian strategic nuclear weapons were mutually detargeted;[16] and in 1998 the alert state of UK *Trident* was relaxed to several days' notice to fire, and has been ever since.[17]

Government studies confirm that a submarine-based missile launching platform is currently the best of a range of options to deliver nuclear weapons.[18] The specific financial cost of ship, submarine and air assets employed to protect the CASD submarine cannot be obtained from MoD sources. Nonetheless, in defence of CASD it is argued that, in circumstances when an SSBN is not on patrol and an escalating threat requires it, the SSBNs are vulnerable to attack in harbour or in transit to deep water; also, the act of deployment exacerbates political tension.

However, it is hard to think of a realistic current scenario in which there is a need to respond to a threat of a 'bolt from the blue' nuclear attack on the UK or other NATO State. Long before circumstances reach the point where nuclear retaliation is the only option, there will be time to deploy an SSBN. Indeed, the act of doing so could be deliberately used as a further essential step up a political ladder of escalation. The Minister of State for the Armed Forces made this very point in evidence to a recent Parliamentary Inquiry on authorising the use of military force.[19] He was referring to 'boots on the ground', but the same logic applies to deploying naval or air assets.

## UK Record on Nuclear Disarmament

The Nuclear Non-Proliferation Treaty (NPT) was signed in 1968 and came into force two years later. Article VI states: "Each of the Parties to the Treaty undertakes to pursue negotiations in good faith on effective measures relating to cessation of the nuclear arms race at an early date and to nuclear disarmament, and on a treaty on general and complete disarmament under strict and effective international control." There have been a number of recent five-yearly NPT reviews where the UK, in lock-step with the US and France, has opposed any measures to include reference to prohibiting and/or reducing its nuclear arsenals. At the conclusion of the May 2019 Preparatory Committee for the 2020 NPT Review Conference, four of the P5 (China was the exception) objected to several recommendations put forward by non-nuclear states such as "the need for a legally-binding norm to prohibit nuclear weapons in order to achieve and maintain a world without nuclear weapons."[20] In consequence they were not adopted.

The Chinese delegation, on the other hand, presented a remarkable and encouraging submission to the Preparatory Committee.[21] It included the following significant statements:
- "Countries possessing the largest nuclear arsenals bear special and primary responsibility for nuclear disarmament and should continue to make drastic and substantive reductions in their nuclear arsenals in a verifiable, irreversible and legally binding manner"; and
- "China undertakes not to be the first to use nuclear weapons at any time and under any circumstances."

The UK, on the other hand, refuses to rule out First Use. The implications of this on *Trident* submarine Commanding Officers is discussed in Part 3 of my book.

The lack of any significant progress in good faith towards the stated NPT goal of complete elimination of nuclear weapons drove 122 non-nuclear Member States of the UN General Assembly to negotiate a Treaty on the Prohibition of Nuclear Weapons (TPNW), which was adopted on 7 July 2017.[22] The NGO 'International Campaign to Abolish Nuclear Weapons' (ICAN) were awarded the 2017 Nobel Peace Prize in recognition of their outstanding work to help generate the political will to achieve this.[23] The TPNW requires ratification by 50 states to come into force.[24] While currently it is most unlikely that any nuclear-armed state will be among them, when the fiftieth state ratifies it, the Treaty's entry into force significantly reinforces the growing international stigmatisation of nuclear deterrence. No doubt this is why the UK boycotted the TPNW negotiations and actively opposes the Treaty.[25]

Since the end of the Cold War the UK has taken the following nuclear disarmament steps:

- After the US and Russia mutually withdrew tactical nuclear weapons from surface ships and submarines in 1991, the UK followed suit a year later. By 1998, all WE.177 free fall nuclear bombs had been withdrawn from the RAF.[26]
- In 1994 PM John Major and Russian President Boris Yeltsin agreed to de-target their deployed strategic nuclear weapons. Subsequently, at the 2000 NPT Review Conference, all the P5 states confirmed that they had mutually de-targeted.[27]
- Reduction to a single nuclear weapon system (*Trident*).
- Reduction to a total of 220 nuclear warheads.
- The deployed SSBN's missiles reduced to eight, with a maximum of 40 warheads.[28]

The last three actions are taken on trust because they are not contained in any form of verifiable international agreement or protocol and so could be reversed at will. By contrast, the basis of US/USSR disarmament negotiations has always been 'trust but verify'.

The UK's 'main gate' decision to go ahead with the *Dreadnought* programme and new warhead ignores the disarmament obligation contained in Article VI of the NPT. It also sends a very hypocritical signal to (for example) North Korea: "We can be trusted to own and responsibly self-regulate our nuclear weapons as a deterrent, but you cannot."

## Summary

The concept of an 'independent nuclear deterrent' is a political chimera. The US has the means, if it so wishes, to prevent the UK using its *Trident* weapon system. The financial and operational costs of sustaining *Trident* and the *Dreadnought* programme are unacceptably threatening the RN's historic core role of graduated conventional deterrence. UK *Trident* missiles have been detargeted since 1994; and since 1998 the deployed SSBN has been at a relaxed notice to fire of several days. With no realistic scenario of a 'bolt from the blue' nuclear threat, there is therefore no justification for maintaining CASD.

For over 20 years now, the UK has failed to pursue significant nuclear disarmament in good faith and has opposed the efforts of other states seeking to ban nuclear weapons. On the contrary, it is modernising its nuclear arsenal and delivery system. Unlike China, it keeps open the option to threaten first use of nuclear weapons, with serious implications for the SSBN command teams.

## Notes

1. HM Government Policy Paper, The UK's nuclear deterrent: what you need to know, 19 February 2018, accessed at https://www.gov.uk/government/publications/uk-nuclear-deterrence-factsheet/uk-nuclear-deterrence-what-you-need-to-know%20

2. British American Security Council, *Trident Commission Final Report*, July 2014, accessed at https://www.basicint.org/

3. Dr Dan Plesch with John Ainslie, Trident: Strategic Dependence & Sovereignty (SOAS, University of London, 2016), accessed at https://www.soas.ac.uk/cisd/news/file114165.pdf

4. Nuclear Information Service (NIS) Report, *AWE: Britain's Nuclear Weapon Factory – Past, Present and Possibilities for the Future*, June 2016, https://nuclearinfo.org/sites/default/files/AWE-Past%2C%20Present%2C%20Future%20Report%202016.pdf

5. Peter Hitchens, Daily Mail, 15 June 2014, accessed at https://www.dailymail.co.uk/debate/article-2657873/Uncovered-American-duplicity-finally-explodes-myth-Special-Relationship-How-US-discussed-blasting-hell-UK-forces-Suez-Crisis-shameful-betrayals-historic-alliance.html

6. Tom Belger, 'U.S secretary of state Pompeo threatens Britain over Huawei 5G row', *Yahoo Finance UK*, 9 May 2019, accessed at https://uk.finance.yahoo.com/news/pompeo-threatens-to-stop-sharing-intelligence-with-uk-over-huawei-5-g-row-144154421.html?guce_referrer=aHR0cHM6Ly93d3cu Z29vZ2xlLmNvbS8&guce_referrer_sig=AQAAAEbMryPunnqF6BzhBLu7r0Nl0ViX50r o-LLuvL5_s5modkpLv0U5bh-80cVdCkzRs0ynNrZuasiEd54Tbnfe25ithpD7Y OkVvnUcmEP0QRsl4UUFAbsBo4EzX0Zx1PuxRu3Wbp8szznTXyRwqKbvmNvH82EN N0k9nWJOh0LO7Tw4&_guc_consent_skip=1597658962

7. Tony Blair, *A Journey* (Hutchinson, London, 2010), p.636.

8. Jana Wattenberg, Aberystwyth University, presentation at RUSI PONI Annual Conference, 13 June 2019, , starting at 00.38, accessed at https://www.nuclearreactions.rusi.org/single-post/2019/07/30/UK-PONI-Conference-2019-A-Summary

9. British American Security Council Report, Blowing up the Budget, 2018, accessed at https://basicint.org/report-blowing-up-the-budget/

10. Admiral Sandy Woodward, *One Hundred days* (Harper Collins, London, 1992), Epilogue, p.348.

11. Exclaimed in frustration after the battle of the Nile, Aug. 1, 1798, being unable to pursue the enemy for want of frigates.

12. Nuclear Information Service Report, *Trouble Ahead*, 29 April 2019, accessed at https://basicint.org/article/nis-reports/new-report-trouble-ahead

13. Save the Royal Navy, *Has the time come to the move the cost of Trident replacement out of the MoD budget*, 27 November 2017, accessed at https://basicint.org/report-blowing-up-the-budget/

14. Commander Robert Green, Royal Navy (Ret'd), *Security without Nuclear Deterrence* (Spokesman Books, Nottingham, 2018) Foreword p.17.

15. HM Government Policy Paper, *The UK's nuclear deterrent: what you need to know*, 19 February 2018, accessed at https://www.gov.uk/government/publications/uk-nuclear-deterrence-factsheet/uk-nuclear-deterrence-what-you-need-to-know

16. Christopher Bellamy, 'Britain stops pointing its missiles at Russia', Independent, 3 June 1994.

17. Select Committee on Defence Eighth Report 20 June 2006, 2. *The UK's Strategic Nuclear Deterrent.* The 1998 Strategic Defence Review, para. 40.

18. HM Government, *Trident Alternatives Review*, 16 July 2013, Executive Summary, para. 32, accessed at https://www.gov.uk/government/publications/trident-alternatives-review

19. RT Hon Mark Lancaster, Minister for the Armed Forces, Oral Evidence to PACAC Inquiry Authorising the use of Military Force, 20 May 2019, in answer to Q.243, accessed at http://data.parliament.uk/writtenevidence/committeeevidence.svc/evidencedocument/public-administration-and-constitutional-affairs-committee/the-role-of-parliament-in-the-uk-constitution-authorising-the-use-of-military-force/oral/102462.pdf

20. Dr Joseph Gerson, 'Human Survival Lies in the Balance as the Charade of NPT Diplomacy is Wearing Thin', In Depth News, 16 May 2019, accessed at https://www.indepthnews.net/index.php/armaments/nuclear-weapons/2689-human-survival-lies-in-the-balance-as-the-charade-of-npt-diplomacy-is-wearing-thin

21. Statement by Chinese Delegation at the Third Session of the Preparatory Committee for the 2020 NPT Review Conference on Nuclear Disarmament, 2 May 2019.

22. UN Office for Disarmament Affairs, Treaty on the Prohibition of Nuclear Weapons 2017.

23. ICAN Nobel Peace Prize, 6 October 2017, see https://icanw.org/

24. ICAN: Status of the Treaty to Prohibit Nuclear Weapons, see https://icanw.org/

25. UK statement on treaty prohibiting nuclear weapons, 8 July 2017, accessed at https://www.gov.uk/government/news/uk-statement-on-treaty-prohibiting-nuclear-weapons%20

26. WE.177, Wikipedia, accessed at https://en.wikipedia.org/wiki/WE.177

27. Statement by the delegations of France, the People's Republic of China, the Russian Federation, the United Kingdom of Great Britain and Northern Ireland and the United States of America, Review Conference of the Parties to the Treaty on the Non-Proliferation of Nuclear Weapons, NPT/CONF/2000/21, 1 May 2000.

28. HM Government Policy Paper, The UK's nuclear deterrent: what you need to know, 19 February 2018, accessed at https://www.gov.uk/government/publications/ uk-nuclear-deterrence-factsheet/uk-nuclear-deterrence-what-you-need-to-know

# WHY TRIDENT?

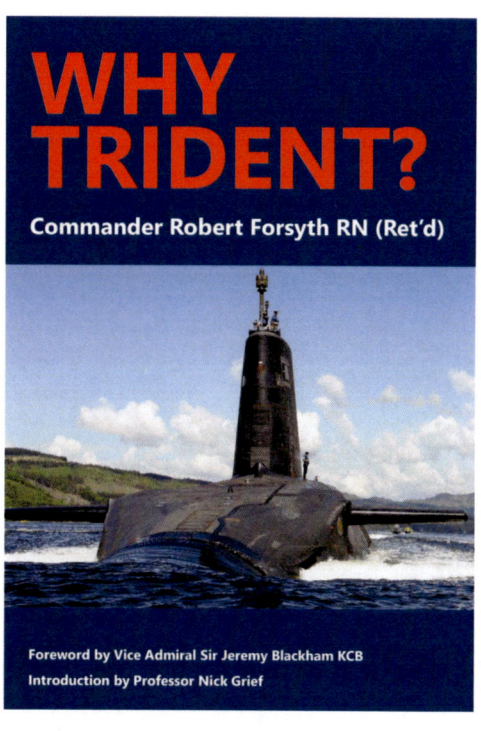

**By Commander Robert Forsyth**
Foreword by Vice Admiral Sir Jeremy Blackham. Introduction by Professor Nick Grief.

£7.99 | 108 Pages | Paperback

"The theory of nuclear deterrence is flawed, unproven and poses significant dangers from accidental use."
Cdr Forsyth

"Few people have done more to examine rigorously the related issues than Rob Forsyth and I am very glad that his work in this field is now being given the exposure it deserves. He has produced a book that all those involved in this field should read and carefully reflect upon."
Vice Admiral Sir Jeremy Blackham

# www.spokesmanbookshop.com

# EUR☮PE

News, updates and information

European Nuclear Disarmament

**END Info** is a regular newsletter produced by the Bertrand Russell Peace Foundation covering disarmament, nuclear threats and anti-nuclear campaigns from a European perspective . . .

**Dowload from**
www.spokesmanbooks.com
www.endinfo.net | @ENDInfo_

**For more information or to subscribe, contact:**
tomunterrainer@russfound.org

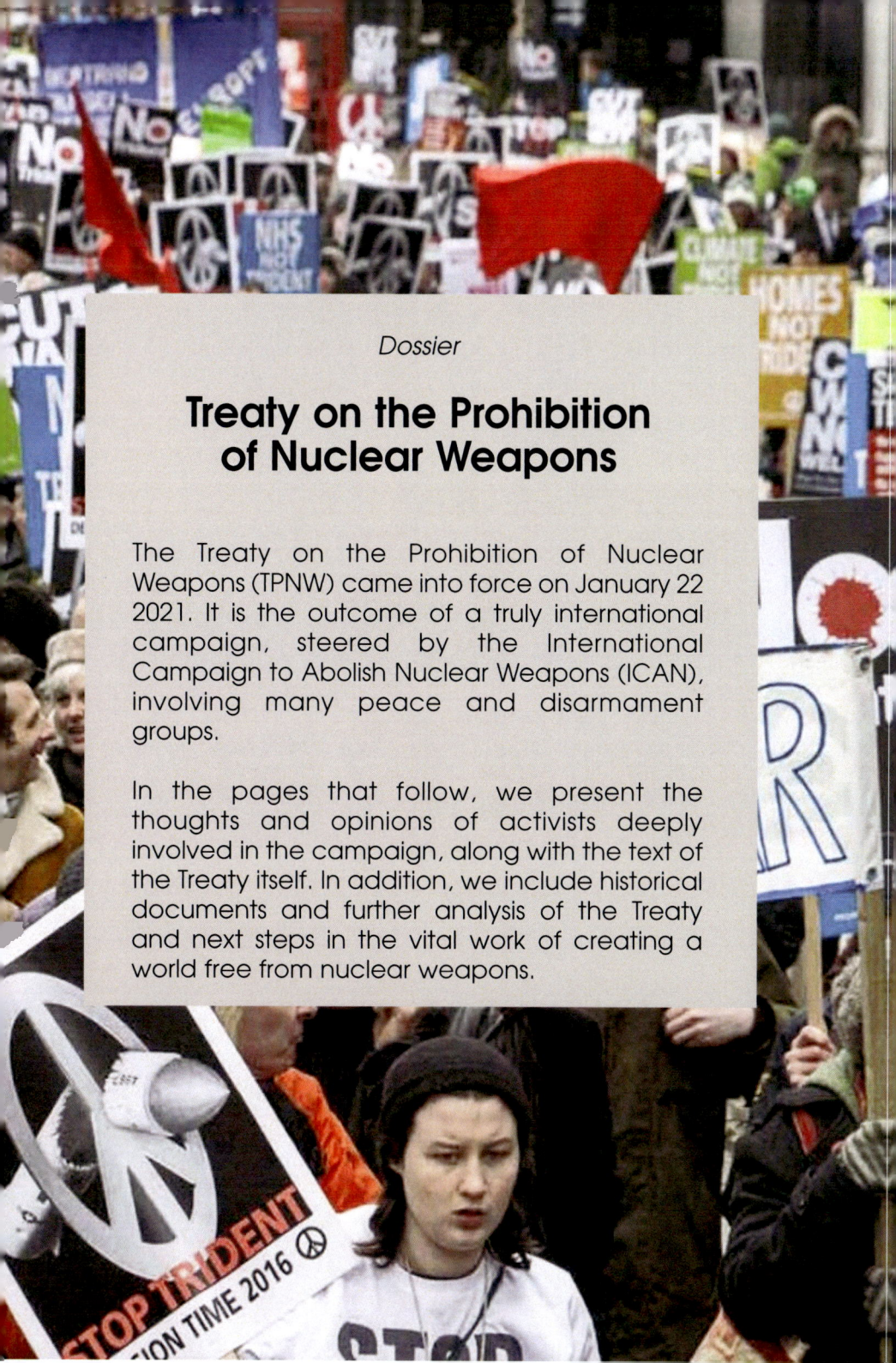

*Dossier*

# Treaty on the Prohibition of Nuclear Weapons

The Treaty on the Prohibition of Nuclear Weapons (TPNW) came into force on January 22 2021. It is the outcome of a truly international campaign, steered by the International Campaign to Abolish Nuclear Weapons (ICAN), involving many peace and disarmament groups.

In the pages that follow, we present the thoughts and opinions of activists deeply involved in the campaign, along with the text of the Treaty itself. In addition, we include historical documents and further analysis of the Treaty and next steps in the vital work of creating a world free from nuclear weapons.

# Through the doorway

*Setsuko Thurlow*

*Setsuko Thurlow survived the atomic bombing of Hiroshima in 1945 and, as she explains, has long campaigned for the abolition of nuclear weapons.*

The Treaty on the Prohibition of Nuclear Weapons has entered into force! This truly marks the beginning of the end of nuclear weapons. When I learned that we reached our 50th ratification, I was not able to stand. I remained in my chair and put my head in my hands and I cried tears of joy. I have committed my life to the abolition of nuclear weapons. I have nothing but gratitude for all who have worked for the success of our treaty. I have a powerful feeling of solidarity with tens of thousands of people across the world. We have made it to this point.

As I sat in my chair, I found myself speaking with the spirits of hundreds of thousands of people who lost their lives in Hiroshima and Nagasaki. I was immediately in conversation with these beloved souls – my sister, my nephew Eiji, other dear family members, my classmates, all the children and innocent people who perished. I was reporting to the dead, sharing this good news first with them, because they paid the ultimate price with their precious lives. Like many survivors, I made a vow that their deaths would not be in vain and to warn the world about the danger of nuclear weapons, to make sure that no one else suffers as we have suffered. I made a vow to work for nuclear disarmament until my last breath. And now we have reached a milestone in our decades' long struggle – the Treaty on the Prohibition of Nuclear Weapons will become international law.

I have a tremendous sense of accomplishment and fulfilment, a sense of satisfaction and gratitude. I know other survivors share these emotions – whether we are survivors from Hiroshima and

Nagasaki; or test survivors from South Pacific island nations, Kazakhstan, Australia and Algeria; or survivors from uranium mining in Canada, the United States or the Congo. All those who have been victimized by the barbaric behavior of nine nations who continue to develop more horrendous weapons, prepared to repeat nuclear massacres far more devastating than the atomic bomb that levelled my hometown, Hiroshima. For the victims and survivors, this initial success with the Treaty on the Prohibition of Nuclear Weapons is tremendously uplifting. I celebrate this moment with my brothers and sisters across the world who have been victimized, and still raise their voices, and still survive.

We also celebrate with those people across the world who recognize the ultimate evil of nuclear weapons, instruments of radioactive violence and omnicide that have kept the entire world hostage for all these 75 years. We celebrate with the global community of anti-nuclear activists who have come together and have worked for the success of this treaty. I am especially grateful to my dear colleagues in the International Campaign to Abolish Nuclear Weapons. ICAN laid the groundwork to collaborate across the divide of diplomacy and activism, to achieve something of profound and lasting importance.

I also want to acknowledge how moved I am that in the preamble to the treaty, *hibakusha* are identified by name. This is the first time in international law that we have been so recognized. We share this recognition with other *hibakusha* across the world, those who have suffered radioactive harm from nuclear testing, from uranium mining, from secret experimentation. And furthermore, the treaty recognizes that indigenous people have been disproportionately affected by the production of nuclear weapons. We in *hibakusha* and indigenous communities uniquely understand that not only the use of nuclear weapons in war but also the testing and production of nuclear weapons causes death and unspeakable suffering through invisible radioactive contamination. And here the treaty recognizes that women and girls are more susceptible to the effects of radiation – that there are gendered implications to radioactive violence.

I am moved to acknowledge the positive obligations of the treaty as well – such as victims' assistance and environmental remediation which will be a hallmark for taking responsibility for the inter-generational effects of radiation. It is vitally important that we all understand that the nuclear age will continue far beyond the nuclear weapon age. We will need to contain and care for radioactive materials into the far future.

But for now, in this joyous present moment, we can rejoice in making our first move. I cannot truly express with words my feelings of overwhelming gratitude. How we have struggled in spite of being confronted by indifference and ignorance. How we have struggled in spite of being ridiculed by nuclear armed and nuclear dependent states. In spite of that and more, we have made it to this point – nuclear weapons are now illegal under international law.

Nuclear abolitionists everywhere can be incredibly encouraged and empowered by this new legal status. Now, with greater intensity and purpose, we will push forward. While this is a time to celebrate, it is not a time for us to relax. The world is ever more dangerous. Yes, we have made it to this point, but we have a long path to cover until we reach our goal of the total elimination of nuclear weapons.

It is unlikely that I will see that day. It is unlikely that any atomic bomb survivor with their own lived memories will bear witness on that day but with the Treaty on the Prohibition of Nuclear Weapons, we can be certain that that beautiful day will dawn. And on that day, we *hibakusha*, test survivors, indigenous people and others, victims to the inter-generational cruelty of radioactive poison, will be remembered and someone alive at present will report to us. Because of our work, our solidarity, our love for this world, we will be a part of a much greater celebration in spirit, when nuclear disarmament will be achieved as part of a greater movement that encompasses peace, justice, equality and compassion for all.

The Treaty on the Prohibition of Nuclear Weapons has opened a new door, wide. Passing through it we begin a new chapter in our struggle – with a mighty embrace of gratitude from those we have lost, and a heartfelt welcome from those who are yet to come.

The beginning of the end of nuclear weapons has arrived! Let us step through the doorway now!

*www.icanw.org*

# Turning Point

*Daryl G. Kimball*

*Daryl Kimball is the executive director of the Arms Control Association.*

The Treaty on the Prohibition of Nuclear Weapons (TPNW) marks a turning point in the long history of the effort to reduce nuclear risks and to eventually eliminate the 13,000 nuclear weapons that remain today, 90 per cent of which are held by the United States and Russia. On Oct. 24, 2020, Honduras became the 50th country to ratify the treaty, triggering its entry into force 90 days later, on Jan. 22, 2021.

That date will mark the first time since the invention of the atomic bomb that nuclear weapons development, production, possession, use, threat of use, and stationing of another country's nuclear weapons on a state party's national territory are all expressly prohibited in a global treaty. The TPNW's entry into force will arrive almost exactly 75 years after the United Nations General Assembly's (UNGA's) adoption, on Jan. 24, 1946, of its very first resolution, Resolution 1(I), which was to establish a commission to ensure "the elimination from national armaments of atomic weapons and all other major weapons adaptable to mass destruction".

This milestone is the culmination of a decade-long initiative spearheaded by a group of key non-nuclear weapon states and a global coalition of civil society campaigners working through the International Campaign to Abolish Nuclear Weapons (ICAN), recipient of the Nobel Peace Prize in 2017 for its unflagging efforts. Negotiations on the TPNW were the outgrowth of a series of three conferences on "The Humanitarian Impacts of Nuclear Weapons" held in Norway, Mexico, and Austria between 2013 and 2014. The so-called "Humanitarian Initiative" re-focused attention on the catastrophic dangers to all humanity posed by nuclear weapons and led to the initiation of negotiations on the treaty at the UN. The talks were completed in July 2017 at the United Nations in New York by a group of more than 120 non-nuclear-weapon states.

The authors of the TPNW argue that because the use of nuclear weapons would violate international humanitarian law, their possession and use must be prohibited. Since the beginning, the major nuclear powers, particularly the United States, Russia, France, and the UK, have tried to slow the momentum toward the TPNW. They claim their security interests justify the perpetuation of their nuclear deterrence doctrines, which envision the potential use of nuclear weapons on a massive scale.

## Filling a Legal Gap

The TPNW effort was designed to fill a "legal gap" in the global nuclear non-proliferation regime regarding the prohibition of nuclear weapons. The 1968 Non-proliferation Treaty (NPT) did not expressly outlaw nuclear weapons, even though their use would be contrary to the rules of international law applicable in armed conflict.

The NPT obligates non-nuclear weapon states to foreswear nuclear weapons, but it recognized the five original nuclear weapon states — the United States, Russia, the United Kingdom, France, and China — already possessed them. Article VI of the NPT obliges all of its 190 States parties to "pursue negotiations in good faith on effective measures relating to cessation of the nuclear arms race at an early date and to nuclear disarmament, and on a treaty on general and complete disarmament under strict and effective international control". But the NPT does not explicitly ban nuclear weapons, and some nuclear-armed states (India, Israel, and Pakistan) are not members of the NPT. So, while the NPT created the environment and pressure for advances on nuclear disarmament, progress has been intermittent and incomplete.

## A Wake-Up Call Amid Rising Risks of Nuclear War

The TPNW arrives at a time when the risks of nuclear war are rising and as the world's major nuclear armed states are building up their nuclear weapons capabilities. It enters into force at the same time that other key agreements limiting nuclear weapons are being discarded or threatened, and as the major nuclear-armed states are failing to meet their NPT nuclear-disarmament obligations. Public attention, too, is focused on other global threats.

The entry into force of the TPNW is a much-needed wake-up call that has the potential to stimulate further action on disarmament and take us closer to a world without nuclear weapons. By strengthening the international legal structure and political norms against nuclear weapons possession and use, the TPNW further delegitimizes nuclear weapons as instruments of power.

The new treaty also reflects the fact, often forgotten in the United States, that for the majority of the world's states, nuclear weapons — and policies that threaten their use for any reason — are immoral, dangerous, and unsustainable. The TPNW is, quite deliberately, a powerful challenge to the nuclear deterrence policies of the world's nine nuclear-armed states, which dangerously assume that military postures that are designed to threaten the use of nuclear weapons on a massive scale can be perfectly managed, even in a crisis, and will never fail to prevent the outbreak of nuclear war. This is why the United States and the other nuclear-armed states boycotted the negotiations and why they have, for the most part, been very critical of the agreement.

### US Critique Misses the Mark

In September 2020, US officials sent a message through diplomatic channels to a significant number of states urging them not to join the TPNW. The missive even urged states that have signed and ratified the treaty to withdraw their support. The letter, accompanied by a "non-paper" listing US concerns about the treaty, stated that: "Although we recognize your sovereign right to ratify or accede to the Treaty on the Prohibition of Nuclear Weapons (TPNW), we believe that you have made a strategic error and should withdraw your instrument of ratification or accession".

The US letter claims that the five original nuclear powers and all members of the NATO alliance "stand unified in our opposition to the potential repercussions" of the treaty. China, however, issued a more conciliatory view in a statement via Twitter on Oct. 24:

> "China has always been advocating complete prohibition and thorough destruction of nuclear weapons, which is fundamentally in line with purposes of [the treaty]. China will continuously make relentless efforts towards a nuclear-weapon-free world."

The US paper also claims that "the TPNW is dangerously counterproductive to the [nuclear Non-proliferation Treaty (NPT)]." This argument is without serious merit. TPNW negotiators have repeatedly underscored that the new treaty seeks full complementarity between the

NPT and the new agreement.

According to Thomas Hajnoczi, director for disarmament, arms control, and non-proliferation for the Austrian Foreign Ministry, "The TPNW did not create a parallel universe to the traditional one founded on the NPT". In an article published in *The Nonproliferation Review* earlier this year, Hajnoczi argues that "on the contrary, it makes the existing universe of legal instruments around the NPT stronger".

The TPNW's negotiators note that the pact advances the existing NPT safeguards regime by legally obliging its state parties to keep in place any additional safeguards arrangements they have voluntarily agreed to implement with the International Atomic Energy Agency.

The TPNW also reinforces the international norm against nuclear testing established by the 1996 Comprehensive Nuclear Test Ban Treaty (CTBT), which prohibits "any nuclear weapon test explosion or any other nuclear explosion". The CTBT has been signed by 184 states.

The TPNW sets forth, for the first time in a nuclear treaty, obligations of assistance to victims of testing and use of nuclear weapons and of environmental remediation of areas affected by testing and use. This reflects the origins of the treaty and the recognition of the unacceptable suffering and devastation that have resulted from the use of nuclear weapons against the people of Hiroshima and Nagasaki and from more than 2,000 nuclear weapons test explosions detonated around the globe, from New Mexico and Nevada and Alaska, across Russia, in Western China, in Kazakhstan, across the South Pacific, in the aboriginal lands in Australia, in Algeria, in South Asia, and until 2017, by North Korea.

Oddly, the US non-paper argues that the pressure that will be exerted by the TPNW to drive nuclear disarmament "disproportionately impacts democracies and democratic alliances" relative to autocratic states like Russia and China, and therefore puts democracies at a disadvantage. According to this strange logic, leaders of democracies should ignore the views of their publics. In fact, history shows that a strength of democratic governments is that they reflect public will. According to a 2020 poll, conducted in the United States July 2–19 for the Chicago Council on Global Affairs, two-thirds (66 per cent) of respondents believe that "no country should be allowed to have nuclear weapons," which is exactly the vision outlined by the TPNW.

Further US efforts to undermine the TPNW are counterproductive and highly divisive. In an Oct. 25 interview with the South African Broadcasting Corporation, South Africa's ambassador to the UN in Geneva, Nozipho Mxakato-Diseko, said: "The US is scared that it cannot justify morally [or] legally the possession of these weapons. They can't blame us [if] they did not want to negotiate [on the TPNW]. Instead they stood outside the General Assembly picketing, [which is] a ridiculous thing for a member of the Security Council. You negotiate with member states. Leadership is about negotiating."

By "asking States to un-ratify a treaty," she said, there is "a danger to the Non-proliferation Treaty. What's to stop you from un-ratifying the NPT? We cannot pick and choose the bits of multilateralism that we want."

ICAN Executive Director Beatrice Fihn said in a statement, "The 50 countries that ratify this Treaty are showing true leadership in setting a new international norm that nuclear weapons are not just immoral but illegal."

### Looking Ahead

Going forward, the United States and the world's other nuclear-armed states should publicly recognize its arrival as a good faith and constructive effort by the majority of the world's nations to eliminate the danger of these weapons and build the legal framework for their eventual elimination.

Now that the TPNW exists, all states — whether they are opponents, supporters, sceptics, or undecideds on the treaty — need to learn to live with it responsibly and find creative ways to move forward together to press for progress on their common challenge: preventing nuclear conflict and eventually ridding the world of nuclear weapons.

With the TPNW restoring the navigational settings for the long journey ahead, responsible states can start by agreeing to a common plan of action on nonproliferation and disarmament at the once-every-five-years gathering to review implementation of the NPT in 2021. Measures they could adopt include, but are not limited to, action that would lead to:

- freezing the size of existing nuclear arsenals and fissile material stockpiles;

- a multilateral agreement on no first use of nuclear weapons;

- securing the ratifications needed to bring the CTBT into force;

- reviving the US-Russian disarmament process, beginning with a five-year extension of New Strategic Arms Reduction Treaty;

- verifiable limits leading to the removal of all shorter-range nuclear weapons, including the 160 US tactical nuclear weapons in five European NATO countries, which would enable those states to join the TPNW;

• banning the introduction of new and destabilizing types of nuclear weapons; and concluding legally binding commitments to not target or threaten non-nuclear-weapon states; and

• agreeing that, as Presidents Ronald Reagan and Mikhail Gorbachev did in 1985, "a nuclear war cannot be won and must never be fought".

Obviously, the new Treaty on the Prohibition of Nuclear Weapons will not reduce the nuclear danger overnight, but it has already changed the conversation. As Martin Luther King said, "The arc of history is long, but it bends toward justice". Clearly, history doesn't bend toward justice on its own, and not quickly enough. The TPNW is going to bend history in the direction that all nations say they want, which is the eventual elimination of nuclear weapons.

> 'There appears to be a presumption that upon Scottish Independence, the Trident submarine fleet and its incredibly destructive WMDs must simply be handed over to Westminster by Holyrood. That is wrong in international law; if the weapons remain on the territory of Scotland, a sovereign state, it will be for the Scottish Government to dispose of them as it chooses.
> 
> The principle is well established and there is a directly relevant and recent precedent in the nuclear weapons in Ukraine. Following the collapse of the Soviet Union, the highly mobile tactical nuclear weapons were swiftly taken back to Russia but the Trident comparators, the strategic nuclear weapons with their silos and the Tupolev strategic bomber fleet and its weapons, were destroyed, many inside Ukraine itself, following the Budapest Agreement of 1994 between the US, UK, Russia and Ukraine and separate bilateral agreements between Ukraine and France, and Ukraine and China.'
> 
> *craigmurray.org.uk*

# Changing Europe's Calculations

*Beatrice Fihn and Daniel Högsta*

*Beatrice Fihn is Executive Director of the International Campaign to Abolish Nuclear Weapons (ICAN). Daniel Högsta is ICAN's Campaign Coordinator.*

On 22 January 2021, nuclear weapons were placed in the same category as chemical and biological weapons – the other weapons of mass destruction – illegal under international law. On that date, the Treaty on the Prohibition of Nuclear Weapons (TPNW) entered into force and will change the legal and normative landscape around nuclear weapons. This has significant implications for any European governments complicit in the practice of deployment and potential use of nuclear weapons of mass destruction.

## A historic milestone for nuclear disarmament

According to experts such as the *Bulletin of Atomic Scientists*, the United Nations Secretary-General, and numerous scientists, academics, and parliamentarians, the risk of nuclear weapons use is increasing. There is no doubt that any use of nuclear weapons would lead to catastrophic harm. The risk is rising in part because of technological modernisation programmes by all the nuclear-armed states and an increasing reliance on weapons of mass destruction by NATO states and nuclear allied states.

The growing international objections to this dangerous trend have been evident in the groundswell of support for the Treaty on the Prohibition of Nuclear Weapons (TPNW). With its 50th state ratification, the TPNW reached the minimum requirements for becoming international law. In response, the United Nations Secretary-General Antonio Guterres called the treaty "a meaningful commitment towards the total elimination of nuclear weapons", and said he looked forward to helping to facilitate the treaty's progress towards this goal. Peter Maurer, President of the International Committee of the Red Cross (ICRC), called it a "historic moment" and "a victory for humanity", which allows us to envision a future without nuclear weapons as "an actual achievable goal".

## Impact on Europe

On 22 January 2021, states parties' obligations under the treaty were triggered. Three states in Europe — Austria, Ireland and Malta — have joined the TPNW. The fact that they find themselves in the minority in Europe is no surprise: the long-standing peer pressure from within NATO and the United States in particular to hold the line against the ban is, by now, well known. But the Treaty will have an impact on other European states – no matter if they join it or not. With its entry into force, the ban on nuclear weapons will be another step towards developing a norm against possessing nuclear weapons. It will positively influence the behaviour of states that are not party to the treaty, including the so-called 'nuclear umbrella states' (which have agreed to be protected by the nuclear weapons of nuclear-armed states), most of whom are on the European continent.

The potential classification of nuclear weapons as 'controversial weapons' by finance industry observers and investors will have implications for banks, pension funds and investment companies. The Treaty will also prompt more discussion of the prohibition of nuclear weapons in parliament, civil society and the media in states not party. Already, over 1,600 elected representatives have called on their governments to join the TPNW, as have capitals in nuclear-armed states such as Paris and Washington D.C.

As has been documented, the practice of nuclear-sharing allows the political risks from – and reputational costs of – participating in the practice of nuclear deterrence to be spread out. The result is a curious situation in which the possessors of nuclear weapons within NATO assert the legitimacy of their weapons by pointing at the obligation to 'defend their allies', while the non-nuclear weapon member states point to the need to have nuclear weapon states involved in any disarmament process – essentially giving the nuclear weapons states the right to veto how and when any such process should start. The TPNW exposes these states for what they are at the moment: complicit in the legitimation of the most destructive and inhumane weapon ever invented – and standing ready to participate in the annihilation of cities and mass murder of civilians. In the states that host US nuclear weapons — Belgium, Germany, Italy, The

Netherlands, and Turkey — the decision to accept these weapons of mass destruction onto their territory was never put to a vote, and for a long time was never publicly acknowledged by their governments. In the other nuclear umbrella states, the public has been told precious little of exactly how their countries will be involved in any decision to use nuclear weapons. Meanwhile, many of these same states pride themselves on being defenders of humanitarian law and democracy. The TPNW is making this double standard very hard to maintain.

**Growing European support for the TPNW**
Public opinion and growing political support for the TPNW is starting to show that the current European government support for nuclear weapons won't be maintained forever. A similar and ultimately futile effort by the United States to keep NATO allies in line behind the use of cluster bombs was known to have taken place during the process leading to the Convention on Cluster Munitions (CCM). In September 2020, 56 ex-presidents and -ministers from 22 nuclear weapons-complicit states signed an open letter calling on the current leaders of their governments to join the TPNW. 46 of them were from Europe, including two former secretaries-general of NATO. The Belgian government's new coalition agreement marks the first instance a NATO government has positively recognised the TPNW. And while discontent around nuclear sharing has long been bubbling in Germany around nuclear weapons, the SPD's Rolf Mützenich caused a stir year when he voiced his opposition to extending the stationing of US nuclear weapons at Büchel air base. More recently, the Green Party – who are strong contenders to join the next government – included a clear commitment to getting Germany to join the TPNW in their party platform for the 2021 elections.

These gains are modest, but they are significant and will grow over time. With major elections in several European NATO states in 2021, the TPNW is guaranteed to continue to be an issue in building coalition governments. Public opinion polls show support for the TPNW across Europe: 79% of Swedes, 78% of Norwegians, 84% of Finns, 70% of Italians, 68% of Germans, 67% of French, and 64% of Belgians support their governments joining the treaty, along with 75% of Japanese.

There's a right way and a wrong way to react to the momentum of the TPNW. The nuclear-armed states, led by the United States, are conducting a master class in doing it wrong. European NATO allies must do better if they want to be on the right side of history.

## What European governments should do next

So what can European governments do? In the spirit of the Belgian government declaration, they should look to how to use the TPNW to advance nuclear disarmament:

- At the NPT Review Conference in 2021, they should acknowledge that the TPNW is an important development for nuclear disarmament as an implementation of NPT's Article 6, as well as reinforcing the non-proliferation pillar.
- Governments should be transparent about their involvement with nuclear weapons and identify ways in which they are in contravention of the norms embodied by the TPNW. Formal reviews should be conducted which explore the steps that would need to be taken in order to join the TPNW.
- The TPNW's first meeting of states parties will take place within 12 months of entry-into-force. While undertaking the aforementioned review, nuclear weapons-complicit states should constructively participate in the MSP as observers, with a view to joining the Treaty in the future.
- The TPNW continues the tradition of other weapons treaties like the CCM and the Anti-personnel Mine Ban Convention by containing positive obligations on Victim Assistance and Environmental Remediation. States unable to join the TPNW at the moment should, as a minimum, contribute to victim assistance and environmental remediation programmes to help the communities that continue to suffer from the devastating impacts of nuclear weapons testing. The nuclear-armed states and their allies have a special responsibility in this respect.

For all the talk of NATO being a 'nuclear alliance for as long as nuclear weapons exist', this is a question of policy rather than a legal obligation under the alliance's founding treaty. And while the TPNW's existence in international law will be constant, policy on nuclear weapons in individual European states, and even in the alliance itself, has and will continue to fluctuate over time.

The risk of nuclear weapons use is growing, in particular, due to emerging technologies that will have a profound impact on warfare in the future, from cyberattacks to artificial intelligence. It is naive and irresponsible to think that European states can maintain their support for the continued existence and deployment of such profoundly dangerous weapons, on the assumptions that deterrence will last forever and no mistakes will ever be made. Their publics do not believe them, and are demanding change.

*First published on the European Leadership Network website:*
*https://www.europeanleadershipnetwork.org/commentary/nuclear-prohibition-changing-europes-calculations/*

# Treaty on the Prohibition of Nuclear Weapons

*The States Parties to this Treaty,*

*Determined* to contribute to the realization of the purposes and principles of the Charter of the United Nations,

*Deeply concerned* about the catastrophic humanitarian consequences that would result from any use of nuclear weapons, and recognizing the consequent need to completely eliminate such weapons, which remains the only way to guarantee that nuclear weapons are never used again under any circumstances,

*Mindful* of the risks posed by the continued existence of nuclear weapons, including from any nuclear-weapon detonation by accident, miscalculation or design, and emphasizing that these risks concern the security of all humanity, and that all States share the responsibility to prevent any use of nuclear weapons,

*Cognizant* that the catastrophic consequences of nuclear weapons cannot be adequately addressed, transcend national borders, pose grave implications for human survival, the environment, socioeconomic development, the global economy, food security and the health of current and future generations, and have a disproportionate impact on women and girls, including as a result of ionizing radiation,

*Acknowledging* the ethical imperatives for nuclear disarmament and the urgency of achieving and maintaining a nuclear-weapon-free world, which is a global public good of the highest order, serving both national and collective security interests,

*Mindful* of the unacceptable suffering of and harm caused to the victims of the use of nuclear weapons (hibakusha), as well as of those affected by the testing of nuclear weapons,

*Recognizing* the disproportionate impact of nuclear-weapon activities on indigenous peoples,

*Reaffirming* the need for all States at all times to comply with applicable international law, including international humanitarian law and international human rights law,

*Basing themselves* on the principles and rules of international humanitarian law, in particular the principle that the right of parties to an armed conflict to choose methods or means of warfare is not unlimited, the rule of distinction, the prohibition against indiscriminate attacks, the rules on proportionality and precautions in attack, the prohibition on the use of weapons of a nature to cause superfluous injury or unnecessary suffering, and the rules for the protection of the natural environment,

*Considering* that any use of nuclear weapons would be contrary to the rules of international law applicable in armed conflict, in particular the principles and rules of international humanitarian law,

*Reaffirming* that any use of nuclear weapons would also be abhorrent to the principles of humanity and the dictates of public conscience,

*Recalling* that, in accordance with the Charter of the United Nations, States must refrain in their international relations from the threat or use of force against the territorial integrity or political independence of any State, or in any other manner inconsistent with the Purposes of the United Nations, and that the establishment and maintenance of international peace and security are to be promoted with the least diversion for armaments of the world's human and economic resources,

*Recalling* also the first resolution of the General Assembly of the United Nations, adopted on 24 January 1946, and subsequent resolutions which call for the elimination of nuclear weapons,

*Concerned* by the slow pace of nuclear disarmament, the continued reliance on nuclear weapons in military and security concepts, doctrines and policies, and the waste of economic and human resources on programmes for the production, maintenance and modernization of nuclear weapons,

*Recognizing* that a legally binding prohibition of nuclear weapons constitutes an important contribution towards the achievement and maintenance of a world free of nuclear weapons, including the irreversible, verifiable and transparent elimination of nuclear weapons, and determined to act towards that end,

*Determined* to act with a view to achieving effective progress towards general and complete disarmament under strict and effective international control,

*Reaffirming* that there exists an obligation to pursue in good faith and bring to a conclusion negotiations leading to nuclear disarmament in all its aspects under strict and effective international control,

*Reaffirming* also that the full and effective implementation of the Treaty on the Non-Proliferation of Nuclear Weapons, which serves as the cornerstone of the nuclear disarmament and non-proliferation regime, has a vital role to play in promoting international peace and security,

*Recognizing* the vital importance of the Comprehensive Nuclear-Test-Ban Treaty and its verification regime as a core element of the nuclear disarmament and non-proliferation regime,

*Reaffirming* the conviction that the establishment of the internationally recognized nuclear-weapon-free zones on the basis of arrangements freely arrived at among the States of the region concerned enhances global and regional peace and security, strengthens the nuclear non-proliferation regime and contributes towards realizing the objective of nuclear disarmament,

*Emphasizing* that nothing in this Treaty shall be interpreted as affecting the inalienable right of its States Parties to develop research, production and use of nuclear energy for peaceful purposes without discrimination,

*Recognizing* that the equal, full and effective participation of both women and men is an essential factor for the promotion and attainment of sustainable peace and security, and committed to supporting and strengthening the effective participation of women in nuclear disarmament,

*Recognizing* also the importance of peace and disarmament education in all its aspects and of raising awareness of the risks and consequences of nuclear weapons for current and future generations, and committed to the dissemination of the principles and norms of this Treaty,

*Stressing* the role of public conscience in the furthering of the principles of humanity as evidenced by the call for the total elimination of nuclear weapons, and recognizing the efforts to that end undertaken by the United Nations, the International Red Cross and Red Crescent Movement, other international and regional organizations, non-governmental organizations, religious leaders, parliamentarians, academics and the hibakusha,

*Have agreed* as follows:

## Article 1: Prohibitions
1. Each State Party undertakes never under any circumstances to:
   (a) Develop, test, produce, manufacture, otherwise acquire, possess or stockpile nuclear weapons or other nuclear explosive devices;
   (b) Transfer to any recipient whatsoever nuclear weapons or other nuclear explosive devices or control over such weapons or explosive devices directly or indirectly;
   (c) Receive the transfer of or control over nuclear weapons or other nuclear explosive devices directly or indirectly;
   (d) Use or threaten to use nuclear weapons or other nuclear explosive devices;
   (e) Assist, encourage or induce, in any way, anyone to engage in any activity prohibited to a State Party under this Treaty;
   (f) Seek or receive any assistance, in any way, from anyone to engage in any activity prohibited to a State Party under this Treaty;
   (g) Allow any stationing, installation or deployment of any nuclear weapons or other nuclear explosive devices in its territory or at any place under its jurisdiction or control.

## Article 2: Declarations
1. Each State Party shall submit to the Secretary-General of the United Nations, not later than 30 days after this Treaty enters into force for that State Party, a declaration in which it shall:
   (a) Declare whether it owned, possessed or controlled nuclear weapons or nuclear explosive devices and eliminated its nuclear-weapon programme, including the elimination or irreversible conversion of all nuclear-weapons-related facilities, prior to the entry into force of this Treaty for that State Party;
   (b) Notwithstanding Article 1 (a), declare whether it owns, possesses or controls any nuclear weapons or other nuclear explosive devices;
   (c) Notwithstanding Article 1 (g), declare whether there are any nuclear weapons or other nuclear explosive devices in its territory or in any place under its jurisdiction or control that are owned, possessed or controlled by another State.

2. The Secretary-General of the United Nations shall transmit all such declarations received to the States Parties.

**Article 3: Safeguards**
1. Each State Party to which Article 4, paragraph 1 or 2, does not apply shall, at a minimum, maintain its International Atomic Energy Agency safeguards obligations in force at the time of entry into force of this Treaty, without prejudice to any additional relevant instruments that it may adopt in the future.

2. Each State Party to which Article 4, paragraph 1 or 2, does not apply that has not yet done so shall conclude with the International Atomic Energy Agency and bring into force a comprehensive safeguards agreement (INFCIRC/153 (Corrected)). Negotiation of such agreement shall commence within 180 days from the entry into force of this Treaty for that State Party. The agreement shall enter into force no later than 18 months from the entry into force of this Treaty for that State Party. Each State Party shall thereafter maintain such obligations, without prejudice to any additional relevant instruments that it may adopt in the future.

**Article 4: Towards the total elimination of nuclear weapons**
1. Each State Party that after 7 July 2017 owned, possessed or controlled nuclear weapons or other nuclear explosive devices and eliminated its nuclear-weapon programme, including the elimination or irreversible conversion of all nuclear- weapons-related facilities, prior to the entry into force of this Treaty for it, shall cooperate with the competent international authority designated pursuant to paragraph 6 of this Article for the purpose of verifying the irreversible elimination of its nuclear-weapon programme. The competent international authority shall report to the States Parties. Such a State Party shall conclude a safeguards agreement with the International Atomic Energy Agency sufficient to provide credible assurance of the non-diversion of declared nuclear material from peaceful nuclear activities and of the absence of undeclared nuclear material or activities in that State Party as a whole. Negotiation of such agreement shall commence within 180 days from the entry into force of this Treaty for that State Party. The agreement shall enter into force no later than 18 months from the entry into force of this Treaty for that State Party. That State Party shall thereafter, at a minimum, maintain these safeguards obligations, without prejudice to any additional relevant instruments that it may adopt in the future.

2. Notwithstanding Article 1 (a), each State Party that owns, possesses or controls nuclear weapons or other nuclear explosive devices shall immediately remove them from operational status, and destroy them as soon as possible but not later than a deadline to be determined by the first meeting of States Parties, in accordance with a legally binding, time-bound plan for the verified and irreversible elimination of that State Party's nuclear-weapon programme, including the elimination or irreversible conversion of all nuclear-weapons-related facilities. The State Party, no later than 60 days after the entry into force of this Treaty for that State Party, shall submit this plan to the States Parties or to a competent international authority designated by the States Parties. The plan shall then be negotiated with the competent international authority, which shall submit it to the subsequent meeting of States Parties or review conference, whichever comes first, for approval in accordance with its rules of procedure.

3. A State Party to which paragraph 2 above applies shall conclude a safeguards agreement with the International Atomic Energy Agency sufficient to provide credible assurance of the non-diversion of declared nuclear material from peaceful nuclear activities and of the absence of undeclared nuclear material or activities in the State as a whole. Negotiation of such agreement shall commence no later than the date upon which implementation of the plan referred to in paragraph 2 is completed. The agreement shall enter into force no later than 18 months after the date of initiation of negotiations. That State Party shall thereafter, at a minimum, maintain these safeguards obligations, without prejudice to any additional relevant instruments that it may adopt in the future. Following the entry into force of the agreement referred to in this paragraph, the State Party shall submit to the Secretary-General of the United Nations a final declaration that it has fulfilled its obligations under this Article.

4. Notwithstanding Article 1 (b) and (g), each State Party that has any nuclear weapons or other nuclear explosive devices in its territory or in any place under its jurisdiction or control that are owned, possessed or controlled by another State shall ensure the prompt removal of such weapons, as soon as possible but not later than a deadline to be determined by the first meeting of States Parties. Upon the removal of such weapons or other explosive devices, that State Party shall submit to the Secretary-General of the United Nations a declaration that it has fulfilled its obligations under this Article.

5. Each State Party to which this Article applies shall submit a report to each meeting of States Parties and each review conference on the progress made towards the implementation of its obligations under this Article, until such time as they are fulfilled.

6. The States Parties shall designate a competent international authority or authorities to negotiate and verify the irreversible elimination of nuclear-weapons programmes, including the elimination or irreversible conversion of all nuclear- weapons-related facilities in accordance with paragraphs 1, 2 and 3 of this Article. In the event that such a designation has not been made prior to the entry into force of this Treaty for a State Party to which paragraph 1 or 2 of this Article applies, the Secretary-General of the United Nations shall convene an extraordinary meeting of States Parties to take any decisions that may be required.

**Article 5: National implementation**
1. Each State Party shall adopt the necessary measures to implement its obligations under this Treaty.

2. Each State Party shall take all appropriate legal, administrative and other measures, including the imposition of penal sanctions, to prevent and suppress any activity prohibited to a State Party under this Treaty undertaken by persons or on territory under its jurisdiction or control.

**Article 6: Victim assistance and environmental remediation**
1. Each State Party shall, with respect to individuals under its jurisdiction who are affected by the use or testing of nuclear weapons, in accordance with applicable international humanitarian and human rights law, adequately provide age- and gender-sensitive assistance, without discrimination, including medical care, rehabilitation and psychological support, as well as provide for their social and economic inclusion.

2. Each State Party, with respect to areas under its jurisdiction or control contaminated as a result of activities related to the testing or use of nuclear weapons or other nuclear explosive devices, shall take necessary and appropriate measures towards the environmental remediation of areas so contaminated.

3. The obligations under paragraphs 1 and 2 above shall be without prejudice to the duties and obligations of any other States under international law or bilateral agreements.

## Article 7: International cooperation and assistance

1. Each State Party shall cooperate with other States Parties to facilitate the implementation of this Treaty.

2. In fulfilling its obligations under this Treaty, each State Party shall have the right to seek and receive assistance, where feasible, from other States Parties.

3. Each State Party in a position to do so shall provide technical, material and financial assistance to States Parties affected by nuclear-weapons use or testing, to further the implementation of this Treaty.

4. Each State Party in a position to do so shall provide assistance for the victims of the use or testing of nuclear weapons or other nuclear explosive devices.

5. Assistance under this Article may be provided, inter alia, through the United Nations system, international, regional or national organizations or institutions, non-governmental organizations or institutions, the International Committee of the Red Cross, the International Federation of Red Cross and Red Crescent Societies, or national Red Cross and Red Crescent Societies, or on a bilateral basis.

6. Without prejudice to any other duty or obligation that it may have under international law, a State Party that has used or tested nuclear weapons or any other nuclear explosive devices shall have a responsibility to provide adequate assistance to affected States Parties, for the purpose of victim assistance and environmental remediation.

## Article 8: Meeting of States Parties

1. The States Parties shall meet regularly in order to consider and, where necessary, take decisions in respect of any matter with regard to the application or implementation of this Treaty, in accordance with its relevant provisions, and on further measures for nuclear disarmament, including:
   (a) The implementation and status of this Treaty;
   (b) Measures for the verified, time-bound and irreversible elimination of nuclear-weapon programmes, including additional protocols to this Treaty;
   (c) Any other matters pursuant to and consistent with the provisions of this Treaty.

2. The first meeting of States Parties shall be convened by the Secretary-General of the United Nations within one year of the entry into force of this Treaty. Further meetings of States Parties shall be convened by the Secretary-General of the United Nations on a biennial basis, unless otherwise agreed by the States Parties. The meeting of States Parties shall adopt its rules of procedure at its first session. Pending their adoption, the rules of procedure of the United Nations conference to negotiate a legally binding instrument to prohibit nuclear weapons, leading towards their total elimination, shall apply.

3. Extraordinary meetings of States Parties shall be convened, as may be deemed necessary, by the Secretary-General of the United Nations, at the written request of any State Party provided that this request is supported by at least one third of the States Parties.

4. After a period of five years following the entry into force of this Treaty, the Secretary-General of the United Nations shall convene a conference to review the operation of the Treaty and the progress in achieving the purposes of the Treaty. The Secretary-General of the United Nations shall convene further review conferences at intervals of six years with the same objective, unless otherwise agreed by the States Parties.

5. States not party to this Treaty, as well as the relevant entities of the United Nations system, other relevant international organizations or institutions, regional organizations, the International Committee of the Red Cross, the International Federation of Red Cross and Red Crescent Societies and relevant non-governmental organizations, shall be invited to attend the meetings of States Parties and the review conferences as observers.

**Article 9: Costs**
1. The costs of the meetings of States Parties, the review conferences and the extraordinary meetings of States Parties shall be borne by the States Parties and States not party to this Treaty participating therein as observers, in accordance with the United Nations scale of assessment adjusted appropriately.

2. The costs incurred by the Secretary-General of the United Nations in the circulation of declarations under Article 2, reports under Article 4 and proposed amendments under Article 10 of this Treaty shall be borne by the

States Parties in accordance with the United Nations scale of assessment adjusted appropriately.

3. The cost related to the implementation of verification measures required under Article 4 as well as the costs related to the destruction of nuclear weapons or other nuclear explosive devices, and the elimination of nuclear-weapon programmes, including the elimination or conversion of all nuclear-weapons-related facilities, should be borne by the States Parties to which they apply.

**Article 10: Amendments**
1. At any time after the entry into force of this Treaty, any State Party may propose amendments to the Treaty. The text of a proposed amendment shall be communicated to the Secretary-General of the United Nations, who shall circulate it to all States Parties and shall seek their views on whether to consider the proposal. If a majority of the States Parties notify the Secretary-General of the United Nations no later than 90 days after its circulation that they support further consideration of the proposal, the proposal shall be considered at the next meeting of States Parties or review conference, whichever comes first.

2. A meeting of States Parties or a review conference may agree upon amendments which shall be adopted by a positive vote of a majority of two thirds of the States Parties. The Depositary shall communicate any adopted amendment to all States Parties.

3. The amendment shall enter into force for each State Party that deposits its instrument of ratification or acceptance of the amendment 90 days following the deposit of such instruments of ratification or acceptance by a majority of the States Parties at the time of adoption. Thereafter, it shall enter into force for any other State Party 90 days following the deposit of its instrument of ratification or acceptance of the amendment.

**Article 11: Settlement of disputes**
1. When a dispute arises between two or more States Parties relating to the interpretation or application of this Treaty, the parties concerned shall consult together with a view to the settlement of the dispute by negotiation or by other peaceful means of the parties' choice in accordance with Article 33 of the Charter of the United Nations.

2. The meeting of States Parties may contribute to the settlement of the

dispute, including by offering its good offices, calling upon the States Parties concerned to start the settlement procedure of their choice and recommending a time limit for any agreed procedure, in accordance with the relevant provisions of this Treaty and the Charter of the United Nations.

## Article 12: Universality
Each State Party shall encourage States not party to this Treaty to sign, ratify, accept, approve or accede to the Treaty, with the goal of universal adherence of all States to the Treaty.

## Article 13: Signature
This Treaty shall be open for signature to all States at United Nations Headquarters in New York as from 20 September 2017.

## Article 14: Ratification, acceptance, approval or accession
This Treaty shall be subject to ratification, acceptance or approval by signatory States. The Treaty shall be open for accession.

## Article 15: Entry into force
1. This Treaty shall enter into force 90 days after the fiftieth instrument of ratification, acceptance, approval or accession has been deposited.

2. For any State that deposits its instrument of ratification, acceptance, approval or accession after the date of the deposit of the fiftieth instrument of ratification, acceptance, approval or accession, this Treaty shall enter into force 90 days after the date on which that State has deposited its instrument of ratification, acceptance, approval or accession.

## Article 16: Reservations

The Articles of this Treaty shall not be subject to reservations.

## Article 17: Duration and withdrawal
1. This Treaty shall be of unlimited duration.

2. Each State Party shall, in exercising its national sovereignty, have the right to withdraw from this Treaty if it decides that extraordinary events related to the subject matter of the Treaty have jeopardized the supreme interests of its country. It shall give notice of such withdrawal to the

Depositary. Such notice shall include a statement of the extraordinary events that it regards as having jeopardized its supreme interests.

3. Such withdrawal shall only take effect 12 months after the date of the receipt of the notification of withdrawal by the Depositary. If, however, on the expiry of that 12-month period, the withdrawing State Party is a party to an armed conflict, the State Party shall continue to be bound by the obligations of this Treaty and of any additional protocols until it is no longer party to an armed conflict.

**Article 18: Relationship with other agreements**
The implementation of this Treaty shall not prejudice obligations undertaken by States Parties with regard to existing international agreements, to which they are party, where those obligations are consistent with the Treaty.

**Article 19: Depositary**
The Secretary-General of the United Nations is hereby designated as the Depositary of this Treaty.

**Article 20: Authentic texts**
The Arabic, Chinese, English, French, Russian and Spanish texts of this Treaty shall be equally authentic.

DONE at New York, this seventh day of July, two thousand and seventeen.

\* \* \*

## Ratification Statements

### Ireland

*Statement by Minister Coveney on the 50th Ratification of the Treaty on the Prohibition of Nuclear Weapons*
I am pleased that the 50th instrument of ratification of the Treaty on the Prohibition of Nuclear Weapons was deposited yesterday, meaning the Treaty will enter into force on 22 January 2021. At a time of rising international tensions, and as we see renewed concerns about nuclear weapons proliferation, a renewed arms race and the destabilising effects of technological developments, the support for the Treaty is a clear indication of the will of the majority of countries to add fresh momentum to achieve

the goal of a world free of nuclear weapons. The significance of the Treaty lies in the fact that for the first time, the core objective of the prohibition of nuclear weapons will be clearly and unambiguously stated in an international Treaty. It challenges us to think about the enormity of the threat posed by these weapons, and by stigmatizing and prohibiting nuclear weapons, it makes a statement that these weapons are no longer acceptable. I am pleased that Ireland ratified the Treaty earlier this year, on the 75th anniversary of the bombing of Hiroshima, continuing our long history of leadership in nuclear disarmament. On this occasion, I pay particular tribute to all victims and survivors of nuclear weapons use and nuclear weapons testing. We owe a debt of gratitude to the survivors who have spent decades campaigning for the elimination of nuclear weapons.

## South Africa

South Africa welcomes the 50th ratification on 24 October 2020 of the UN Treaty on the Prohibition of Nuclear Weapons (TPNW) allowing it to enter into force. The seminal Treaty was adopted by United Nations member states on 07 July 2017 and South Africa signed the Treaty at a signing ceremony held on the margins of the 72nd Session of the United Nations General Assembly in September 2017 and ratified it on 25 February 2019.

The Treaty prohibits all signatory countries from developing, testing, producing, manufacturing, transferring, possessing, stockpiling, using or threatening to use nuclear weapons, or allowing nuclear weapons to be stationed on their territory. It also prohibits them from assisting, encouraging or inducing anyone to engage in any of these activities.

Welcoming the final step for the Treaty to come into force, International Relations and Cooperation Minister, Naledi Pandor stated ... 'South Africa is honoured, as the first country to have voluntarily eliminated all its nuclear weapons, to have played a leading role, together with several UN member states and members of civil society in ensuring that the Treaty is agreed upon and now finally ratified. The Treaty exemplifies the central goal of the United Nations, which according to the UN Charter is to "save succeeding generations from the scourge of war'. The Minister concluded, 'Our collective goal must remain to achieve a world free of nuclear weapons. In this regard, South Africa, working with others who maintain the same goal, will continue our efforts to ensure the complete elimination of all weapons of mass destruction'.

The Treaty complements other international instruments by contributing towards fulfilling the nuclear disarmament obligations under the Nuclear Non-Proliferation Treaty (NPT), the objectives of the Comprehensive

Nuclear Test-Ban Treaty (CTBT) and the various nuclear-weapon-free-zone treaties, such as the Pelindaba Treaty that already banned nuclear weapons in Africa.

## Austria

'The rapid entry into force of this central prohibition standard, only three years after I signed the treaty for Austria in New York, is also a success of our close cooperation with friendly states, the Red Cross and civil society. It is appalling that 75 years after the atomic bombs dropped on Hiroshima and Nagasaki, with their devastating humanitarian consequences, we are still not safe from these despicable weapons. With the entry into force of the Treaty on the Prohibition of Nuclear Weapons, we are making it very clear that we do not accept a standstill in nuclear disarmament and that nuclear deterrence does not create security. It is high time to finally put an end to this myth', stressed Foreign Minister Schallenberg.

The Treaty on the Prohibition of Nuclear Weapons reached the 50 ratifications necessary for its entry into force yesterday, Saturday – just on the 75th anniversary of the United Nations. The treaty will enter into force 90 days later. The Treaty on the Prohibition of Nuclear Weapons was negotiated on Austria's initiative: 122 UN member states approved the text of the treaty on 7 July 2017, and therefore clearly opposed the rearmament programmes of nuclear-armed states – programmes which cost billions of euros.

'In times of rising geopolitical tensions and the modernisation of arsenals, the entry into force of this treaty is a clear signal that disarmament commitments can no longer remain hollow words and that the risks and permanent threat posed by these weapons of mass destruction are unacceptable – and finally illegal, too', explained a convinced Federal Chancellor Kurz and Foreign Minister Schallenberg.

The adoption of the prohibition treaty was preceded by three conferences on the humanitarian consequences of nuclear weapons in close cooperation with the Red Cross and civil society. At the end of the Vienna Conference in 2014, Austria announced the drafting of a ban. This initiative was joined by 127 states, whereupon work began on drafting the treaty. As a further sign of its commitment, Austria has already proposed holding the first meeting of the signatories in Vienna at the headquarters of the United Nations. This should take place within one year of the treaty's entry into force. 'Until then we are calling on all governmental and non-governmental partners to join us in maintaining the pressure for further signatures and ratifications of the treaty, so we can achieve our common

### China

*Tweet sent by the 'Chinese Mission to UN'*
China has always been advocating complete prohibition and thorough destruction of nuclear weapons, which is fundamentally in line with purposes of #TPNW. China will continuously make relentless efforts towards a nuclear-weapon-free world.

---

Resolution 1 (I) was passed on January 24, 1946. It was the very first resolution passed by the United Nation's General Assembly

### 1 (I). ESTABLISHMENT OF A COMMISSION TO DEAL WITH THE PROBLEMS RAISED BY THE DISCOVERY OF ATOMIC ENERGY

*Resolved by the General Assembly of the United Nations* to establish a commission, with the composition and competence set out hereunder, to deal with the problems raised by the discovery of atomic energy and other related maters:

**1. ESTABLISHMENT OF THE COMMISSION**
A Commission is hereby established by the General Assembly with the terms of reference set out under section 5 below …

**5. TERMS OF REFERENCE OF THE COMMISSION**
The Commission shall proceed with the utmost despatch and enquire into all phases of the problem, and make such recommendations from time to time with respect to them as it finds possible. In particular, the Commission shall make specific proposals:
(a) for extending between all nations the exchange of basic scientific information for peaceful ends;
(b) for control of atomic energy to the extent necessary to endure its use only for peaceful purposes;
(c) for the elimination from national armaments of atomic weapons and of all other major weapons adaptable to mass destruction;
(d) for effective safeguards by way of inspection and other means to protect complying States against the hazards of violations and evasions.

goal of a world free from nuclear weapons', concluded Federal Chancellor Kurz and Foreign Minister Scahallenberg.

# Disarmament

*Bertrand Russell*

*An extract from Bertrand Russell's 1959 book,* Common Sense and Nuclear Warfare.

There are many who consider that the problem of agreed disarmament or reduction of armaments is the most important in the field of international relations and the one to be first dealt with. I do not share this view. Needless to say, I consider agreed reduction of armaments very important and I favour the complete prohibition of all nuclear weapons, whether strategic or tactical. I see, however, two objections to treating this as the central and primary problem: First, as the experience of the last thirteen years has shown, disarmament conferences cannot reach agreements until the relations of East and West become less strained than they have been; second, the long-run problem of saving mankind from nuclear extinction will only be postponed, not solved, by agreements to renounce nuclear weapons. Such agreements will not, of themselves, prevent war, and, if a serious war should break out, neither side would consider itself bound by former agreements, and each side would, in all likelihood, set to work to manufacture new H-bombs as quickly as possible. These two considerations belong to different ends of the long road towards secure peace. The first prevents nations from starting along the road; the second shows a possibility of their being deflected after travelling a long way towards the goal. For these reasons, I should regard agreed disarmament as a palliative rather than a solution.

Nevertheless, the importance of any agreed measure of disarmament would be very great indeed. Perhaps its first and greatest importance would consist in the proof that negotiations between East and West can bear fruit in measures that all sane men must welcome. The second gain would be a diminution of the risk of unintended war.

The present readiness for instant retaliation makes it possible for some wholly accidental misfortune, such as a meteor exploding an H-bomb, to be mistaken for enemy action. Since it is assumed, probably rightly, that a Great Power, if embarked upon nuclear war, would begin by destroying the seat of government of the enemy, it is inferred that subordinate commanders must not wait for orders from headquarters but must carry out plans previously arranged to meet the emergency. Many things more probable than collision with a meteor might initiate a war that no Great Power had intended. One such cause would be a mechanical defect in radar. Another would be a sudden nervous breakdown of some important officer as a result of the stress caused by appalling responsibility. A third, and even more likely source of danger, will arise when many countries have nuclear weapons. It will then be possible for a small country with an irresponsible, chauvinistic Government, to make a nuclear attack which would be interpreted as coming from a major Power and would, therefore, lead to world war before the error was discovered. For such reasons, the present state of the world, and still more the state which will exist when, as now seems nearly certain, a great many States possess H-bombs, involves quite appalling dangers which could be very greatly lessened by disarmament agreements.

A third reason for desiring a reduction of armaments is economy. The importance of this reason is likely to increase and become more evident during the next few years. Western Governments, faced by fear of mounting expenditure, have recently adopted the view that nuclear weapons almost alone could afford adequate defence. This view is being increasingly challenged by experts on the ground that the United States could suffer unendurably from a nuclear attack and would, therefore, be very unwilling to provoke a nuclear war. It follows that, if the West is to be capable of resisting the East without disaster, it must be able to conduct non-nuclear wars, although the ability to do so involves enormously increased expenditure. Apart from this somewhat technical consideration, one must assume that, so long as the arms race continues and remains a matter of life and death to both sides, new inventions will constantly increase military expenditure until both sides are reduced to subsistence level. The only escape will be when both sides realize that it is more profitable to keep one's own citizens prosperous than to be able to kill those of other countries.

The fourth gain which may be secured by disarmament agreement is that they may show the necessity of deciding disputes by arbitration or by some international tribunal, rather than by war or the threat of war. This is an almost inevitable logical consequence of any such agreement. Decision by war implies the use of the whole of a nation's strength if that is necessary for victory. A disarmament agreement on the other hand, so long as it is respected, implies that the Government is not using its whole strength in preparation for war. This leads inevitably to the conclusion that new methods of settling disputes must be sought. Granted that a reduction of armaments is desirable, we are faced at once by formidable problems. After studying the proceedings of disarmament conferences, it is almost impossible not to be lost in a morass of technicalities, with arguments this way and that and well-founded objections that are met by equally well-founded retorts. So long as the East-West tension remains what it has been, I do not think that we are likely to escape from this morass. Suppose the East offers to agree to the abolition of all nuclear weapons. The West at once retorts that the superior man-power of the East would give it an unfair advantage unless conventional armaments were reduced at the same time. Suppose this admitted. The next question that arises is: To what figure should the conventional armaments of East and West be reduced? Suppose this agreed, there arises a third and most difficult question: What endurable measures of inspection will insure that an agreement is being loyally carried out? Hitherto it has been found that such questions could be prolonged *ad infinitum* and that negotiators could continue throughout many years to advocate disarmament without incurring the risk of bringing it about. If disarmament negotiations are to succeed, it will only be when each side is persuaded that the other has abandoned the hope of conquest.

There is, it is true, one measure which is already within the sphere of practical politics, and that is the abolition of nuclear tests. What makes this measure already possible is that scientists are agreed in believing that no serious nuclear test can be concealed, given a system of inspection so little onerous that neither side objects to it. Although the stoppage of tests is only a small step, it will nevertheless be very welcome if it takes place. It will be welcome, first, because it will put an end to the increase of radioactive substances in air and water and food which at present is causing an increase of cancer and leukaemia and genetic damage of unknown magnitude. It will be welcome, in the second place, because any agreement between East and West is to the good and tends to diminish tension. It will be welcome, in the third place, because it will make it more difficult for new Powers to join the 'Nuclear Club'. For these reasons, we must all ardently hope that an agreement to abolish tests will be reached.

Apart from the absence of any genuine governmental desire for disarmament, the greatest difficulties are connected with the question of inspection. On this subject there is an admirable book: *Inspection for Disarmament*, edited by Seymour Melman, and published by the Columbia University Press, New York, in 1958. So far as I am able to judge, the investigations contained in this book are completely honest and aim solely at a just estimate of facts and probabilities. Broadly speaking, the conclusion reached in this book is that inspection could prevent the manufacture of new nuclear weapons, but that it probably could not prevent a dishonest Government from concealing some part of the stocks existing at the time when an agreement was concluded. There is a valuable account of the devices by which the German Government, after the First World War, concealed the armaments which it created in defiance of the Treaty of Versailles. In this case, the acquiescence of the German Government in the disarmament clauses of that Treaty was not voluntary, but was only a reluctant acquiescence in the consequences of defeat. I think we may infer that no disarmament agreement will be reliable unless all signatory States are sincerely convinced that it is to their own advantage, and not only to that of potential enemies. This re-enforces our earlier contention that disarmament must result from better relations between East and West, and cannot, by itself, be a cause of such better relations.

Given a genuine desire for peace on both sides, it should be possible, without undue delay, to agree that no new nuclear weapons should be manufactured. This is a measure which could be enforced by inspection without great difficulty. Aerial inspection, especially, would make the concealment of large plants almost impossible, even in the remotest regions of Siberia or Alaska. The destruction of existing stocks of H-bombs should follow, but offers greater difficulties, and, if it is to be carried out without altering the balance of power, it will have to be accompanied by a reduction of conventional forces. I doubt whether an agreement to this effect will be concluded until there is a genuine readiness on both sides to renounce war as an instrument of policy.

I should like, in conclusion, to say a few words about the increase of general well-being that would result if such measures of disarmament as we have been discussing were carried out. I put first among the gains to be expected the removal of that terrible load of fear which weighs at present upon all those who are aware of the dangers with which mankind is threatened. I believe that a great upsurge of joy would occur throughout the civilized world and that a great store of energies now turned to hate and destruction and futile rivalry would be diverted into creative channels, bringing happiness and prosperity to parts of the world which, throughout long ages, have been oppressed by poverty and excessive toil. I believe that the emotions of kindliness, generosity and sympathy, which are now kept within iron fetters by the fear of what enemies may do, would acquire a new life and a new force and a new empire over human behaviour. All this is possible. It needs only that men should permit themselves a life of freedom and hope from which they are now excluded by the domination of unnecessary fear.

# Challenging Nuclearism

*The UN Treaty on the Prohibition of Nuclear Weapons*

*Richard Falk*

*Richard Falk is professor emeritus of international law at Princeton University and was Professor of Global Law, Queen Mary University of London. He served a six-year term as United Nations Special Rapporteur on the situation of human rights in Palestinian territories. He is the author or co-author of numerous books about global governance, human rights, and the idea of world order. He is Senior Vice President at the Nuclear Age Peace Foundation and has acted as counsel before the International Court of Justice.*

On 7 July 2017, 122 countries at the UN voted to approve the text of a proposed international treaty entitled 'Draft Treaty on the Prohibition of Nuclear Weapons.' (TPNW) The treaty was formally opened for signature that September, but it only became a binding legal instrument according to its own provisions on January 21, 2021, which is 90 days after the 50th country deposited with the UN Secretary General its certification that the treaty has been ratified in accordance with their various constitutional processes. This is a major accomplishment, not least because all of the major nuclear weapons states refused to participate in the negotiating process, and the United States, France, and UK issued a formal statement denouncing the treaty and refusing to alter their reliance on nuclear weapons in carrying out their foreign policy

In an important sense, it is incredible that it took 76 years after the attacks on Hiroshima and Nagasaki to reach this point of setting forth an unconditional prohibition of any use or threat of nuclear weapons [Article 1(e)] within the framework of a multilateral treaty negotiated under UN auspices. The core obligation of states that choose to become parties to the treaty is very sweeping. It prohibits any connection whatsoever with the weaponry by way of possession, deployment, testing, transfer, storage, and production [Article 1(a)].

The TPNW is significant beyond the prohibition. It can and should be interpreted as a frontal rejection of the geopolitical approach to nuclearism, and its contention that the retention and development of nuclear weapons is a proven necessity for global security given the way international society is organized. It is a healthy development that the TPNW shows an impatience toward and a distrust of the elaborate geopolitical rationalizations of the nuclear status quo that have ignored the profound objections to nuclearism of many governments and the anti-nuclear views that have long dominated world public opinion and animated civil society activists. The old reassurances of the nuclear weapons states about being committed to nuclear disarmament as soon as an opportune moment arrives increasingly lack credibility as the nuclear weapons states, led by the United States, make continuing huge investments in the modernization and further development of their nuclear arsenals, with the US even proposing to deploy nuclear weapons in space, despite the risks and expense.

Despite this justifiable sense of achievement, it must be admitted that there is a near fatal weakness, or at best, a gaping hole in this newly cast net of prohibition established by way of the TPNW process. True, 122 signatures, and even more, the formal entry into force of the treaty, lends weight to the claim that the international community, by taking such a significant stand has signalled in an obligatory way the repudiation of nuclear weapons for any and all purposes, and formalized the prohibition of any action to the contrary. The enormous fly in this healing ointment arises from the refusal of any of the nine nuclear weapons states to join in the TPNW process even to the legitimating extent of participating in the negotiating conference with the opportunity to express their objections and influence the outcome. As well, most of the chief allies of these states that are part of the global security network of states relying directly and indirectly on nuclear weaponry also boycotted the entire process. It is also discouraging to appreciate that several countries in the past that had lobbied against nuclear weapons with great passion such as India, Japan, and China were notably absent, and also opposed the prohibition. This posture of undisguised opposition to this UN sponsored undertaking to delegitimize nuclearism, while reflecting the views of a minority of governments, must be taken extremely seriously. It includes all five permanent members of the Security Council that have sophisticated nuclear weapons programs of their own, and such important international actors as Germany and Japan that have long taken shelter under the US nuclear umbrella.

The NATO triangle of France, United Kingdom, and the United States, three of the five veto powers in the Security Council, angered by its inability to prevent the whole TPNW venture, went to the extreme of issuing a Joint Statement of denunciation in 2017, the tone of which was disclosed by its defiant assertion removing any doubt as to the abiding commitment to a nuclearized world order: 'We do not intend to sign, ratify or ever become party to it. Therefore, there will be no change in the legal obligations on our countries with respect to nuclear weapons.' The body of the statement contended that global security depended upon maintaining the nuclear status quo, as bolstered by the Non-proliferation Treaty (NPT) of 1968 and by the claim that it was 'the policy of nuclear deterrence, which has been essential to keeping the peace in Europe and North Asia for over 70 years'. It is relevant to take note of the geographic limits associated with the claimed peace-maintaining benefits of nuclear weaponry, which ignores the ugly reality that devastating warfare has raged throughout this period outside the feared mutual destruction of the heartlands of geopolitical rivals, a central shared forbearance by the two nuclear superpowers throughout the entire Cold War. During these decades of rivalry, the violent dimensions of geopolitical rivalry were effectively outsourced to the non-Western regions of the world, and subsequently, causing massive suffering and widespread devastation for many vulnerable peoples inhabiting Asia, Africa, and the Middle East. Such a conclusion suggests that even if we were to accept the claim that nuclear weapons deserve credit for avoiding a major war, specifically World War III, that 'achievement' was accomplished at the cost of millions, probably tens of millions, of civilian lives in non-Western societies. Beyond this, the achievement involved a colossally irresponsible gamble with the human future, succeeded as much due to good luck as to the rationality attributed to deterrence theory and practice, an assessment confirmed in Martin Sherwin's definitive historical study, *Gambling with Armageddon: Nuclear Roulette from Hiroshima to the Cuban Missile Crisis* (2020).

TPNW itself does not challenge the Westphalian framework of state-centrism by setting forth a framework of global legality that is issued under the authority of 'the international community' or the UN as the authoritative representative of the peoples of the world. Its provisions are carefully formulated as imposing obligation only with respect to 'State parties,' that is, governments that have deposited the prescribed ratification and thereby become formal adherents of the treaty. Even Article 4, which hypothetically details how nuclear weapons states should divest themselves of all connections with the weaponry limits its claims to State parties, and offers no guidance whatsoever in the event of suspected or alleged non-compliance. Reliance is placed in Article 5 on a commitment to secure compliance by way of the procedures of 'national implementation.'

The treaty does aspire to gain eventual universality through the adherence of all states over time, but in the interim the obligations imposed are of minimal substantive relevance beyond the agreement of the non-nuclear parties not to accept deployment or other connections with the weaponry. It is for another occasion, but I believe a strong case can be made under present customary international law, emerging global law, and abiding principle of natural law that the prohibitions in the TPNW are binding universally independent of whether a state chooses or not to become a party to the treaty.

Taking an unnecessary further step to reaffirm statism, and specifically, 'national sovereignty' as the foundation of world order, Article 17 gives parties to the TPNW a right of withdrawal. All that state parties have to do is give notice, accompanied by a statement of 'extraordinary circumstances' that have 'jeopardized the supreme interests of its country.' The withdrawal will take effect twelve months after the notice and statement are submitted. There is no procedure in the treaty by which the contention of 'extraordinary circumstances' can be challenged as unreasonable or made in bad faith. It is an acknowledgement that even for these non-nuclear states adhering to the treaty, nothing in law or morality or human wellbeing takes precedence over the exercise of sovereign rights. Article 17 is not likely to be invoked in the foreseeable future. This provision reminds us of this strong residual unwillingness to supersede *national* interests by deference to *global* and *human* interests. The withdrawal option is also important because it confirms that national security continues to take precedence over international law, even with respect to genocidal weaponry of mass destruction. As such the obligation undertaken by parties to the TPNW are reversible in ways that are not present in multilateral conventions outlawing genocide, apartheid, and torture, or in *jus cogens* domains.

Given these shortcomings, is it nevertheless reasonable for nuclear abolitionists to claim a major victory by virtue of tabling such a treaty? Considering that the nuclear weapons states and their allies have all rejected the process and even those within the circle of the intended legal prohibition reserve a right of withdrawal, the TPNW is likely to be brushed aside by realists and cynics as mere wishful thinking and even by some dedicated anti-nuclearists as more of an occasion for hemlock than champagne. The cleavage between the nuclear weapons states and the rest of the world has never been starker, and there are absent any signs on either side of the divide to make the slightest effort to find common ground, and there may be none. As of now, it is a standoff between two forms of asymmetry. The nuclear states enjoy a preponderance of hard power, while the anti-nuclear states have the upper hand when it comes to soft power, including solid roots in 'substantive democracy,' 'global law,' 'natural law,' and 'global ethics.'

The hard power solution to nuclearism has essentially been reflexive, that is, relying on nuclearism as shaped by the leading nuclear weapons states. What this has meant in practice is some degree of self-restraint on the battlefield and crisis situations (there is an existential nuclear taboo without doubt, although it has never been seriously tested), and, above all, a delegitimizing one-sided implementation of the Non-proliferation Treaty regime. This one-sidedness manifests itself in two ways: (1) discriminatory administration of the underlying non-proliferation norm, most unreservedly in the case of Israel; as well, the excessive enforcement of the non-proliferation norm beyond the limits of either the NPT itself or the UN Charter, as with Iraq (2003), and currently by way of threats of military attack against North Korea and Iran. Any such uses of military force would be non-defensive and unlawful unless authorized by a Security Council resolution supported by all five permanent members, and at least four other states, which fortunately remains unlikely. [UN Charter, Article 27(3)] More likely is recourse to unilateral coercion led by the countries that issued the infamous joint declaration denouncing the TPNW as was the case for the US and the UK with regard to recourse to the war against Iraq, principally rationalized as a counter-proliferation undertaking, which turned out itself to be a rather crude pretext for mounting an aggressive war with other goals, showcasing 'shock and awe' tactics.

(2) The failure to respect the obligations imposed on the nuclear weapons states to negotiate in good faith an agreement to eliminate these weapons by verified and prudent means, and beyond this to seek agreement on general and complete disarmament. It should have been evident, almost 50 years after the NPT came into force in 1970 that nuclear weapons states have breached their material obligations under the treaty, which were validated by an Advisory Opinion of the International Court of Justice in 1996 that included a unanimous call for the implementation of these Article VI legal commitments. Drawing this conclusion from deeds as well as words, it is evident for all with eyes that want to see, that the nuclear weapons states as a group have opted for deterrence plus counter-proliferation as their *permanent* security regime.

One contribution of the TPNW is to convey to the world the crucial awareness of these 122 countries as reinforced by global public opinion that the deterrence/NPT approach to global peace and security is neither prudent nor legitimate nor a credible pathway leading over time to the end of nuclearism.

In its place, the TPNW offers its own two-step approach—first, an unconditional stigmatizing of the use or threat of nuclear weapons to be followed by a negotiated process seeking nuclear disarmament. Although the TPNW is silent about demilitarizing geopolitics and conventional disarmament, it is widely assumed that latter stages of denuclearization would not be implemented unless they involved an ambitious downsizing of the war system. The TPNW is also silent about the relevance of nuclear power capabilities, which inevitably over time entail a weapons option given widely available current technological knowhow. The relevance of nuclear energy technology would have to be addressed at some stage of nuclear disarmament.

Having suggested these major shortcomings of treaty coverage and orientation, can we, should we cast aside these limitations, and join in the celebrations and renewed hopes of civil society activists to rid the world of nuclear weapons? My esteemed friend and colleague, David Krieger, who has dedicated his life to keeping the flame of discontent about nuclear weapons burning and serves as the longtime and founding President of the Nuclear Age Peace Foundation, concludes his informed critique of the Joint Statement by NATO leaders with this heartening thought: "Despite the resistance of the US, UK and France, the nuclear ban treaty marks the beginning of the end of the nuclear age." [Krieger, "U.S., UK and France Denounce the Nuclear Ban Treaty"]. I am not at all sure about this, although Krieger's statement leaves open the haunting uncertainty of how long it might take to move from this 'beginning' to the desired 'end.' Is it as some self-styled 'nuclear realists' like to point out, no more than an *ultimate* goal, which is polite coding for the outright dismissal of the nuclear disarmament option as 'utopian' or 'unattainable'?

We should realize that there have been many past 'beginnings of the end' since 1945 that have not led us any closer to the goal of the eliminating the scourge of nuclearism from the face of the earth. It is a long and somewhat arbitrary list, including the immediate horrified reactions of world leaders to the atomic bomb attacks at the end of World War II, and what these attacks suggested about the future of warfare; the massive antinuclear civil disobedience campaigns that briefly grabbed mass attention in several nuclear weapons states; tabled disarmament proposals by the United States and the Soviet Union in the 1960s; the UN General Assembly Resolution 1653 (XVI) that in 1961 declared threat or use of nuclear weapons to be unconditionally unlawful under the UN Charter and viewed any perpetrator as guilty of a crime against humanity; the Cuban Missile Crisis of 1962 that scared many at the momentary realization that it was not tolerable to coexist with nuclear weapons; the International Court of Justice majority Advisory Opinion in 1996 responding to the General Assembly's formal inquiry about the legality of nuclear weapons, limiting the possibility of legality of use to the narrow circumstance of responding to imminent threats to the survival of a sovereign state; the apparent proximity to historic disarmament arrangements agreed to by Ronald Reagan and Mikhail Gorbachev at a summit meeting in Reykjavik, Iceland in 1986; the extraordinary opening provided by the ending of the Cold War and the collapse of the Soviet Union, which offered world leaders the best possible 'beginning of the end,' and yet nothing happened; and finally, Barack Obama's Prague speech is 2009 (echoing sentiments expressed less dramatically by Jimmy Carter in 1977, also early in his presidency) in which he advocated to great acclaim dedicated efforts to advance toward the elimination of nuclear weapons if not in his lifetime, at least as soon as possible; it was a good enough beginning for a Nobel Peace Prize, but then one more fizzle, presumably discouraged by the pushback of the formidable nuclear weapons establishment.

Each of these occasions briefly raised the hopes of humanity for a future freed from a threat of nuclear war, and its assured accompanying catastrophe, and yet there was few, if any, signs of progress flowing from each of these beginnings greeted so hopefully toward the ending posited as a goal. Soon disillusionment, denial, and distraction overwhelmed the hopes raised by these earlier initiatives, with the atmosphere of hope in each instance replaced by an aura of nuclear complacency, typified by indifference, ignorance, and denial. It is important to acknowledge that the national bureaucratic and ideological *structures* supporting nuclearism are extremely resilient, and have proved adept at outwaiting and outwitting the flighty politics of periodic flurries of anti-nuclear activism.

And after a lapse of years, yet another new beginning is now being proclaimed. We need to summon and sustain greater energy than in the past if we are to avoid this fate of earlier new beginnings in relation to the TPNW. We need to do our best to let this latest beginning start a process that moves steadily toward the end that has been affirmed. We know that the TPNW would not itself have moved forward without civil society militancy and perseverance at every stage. The challenge now is to discern and then take the next steps, and not follow the precedents of the past that followed the celebration of a seeming promising beginning with a misplaced reliance on the powers that be to handle the situation, and act accordingly. In the past, the earlier beginnings were soon buried, acute concerns eventually resurfaced, and yet another new beginning was announced with fanfare while the earlier failed beginnings were purged from collective memory.

Here, we can at least thank the Joint Statement of leading NATO allies for sending a clear signal to civil society and the 122 governments voting their approval of the TPNW text that if they are truly serious about ending nuclearism, they will have to carry on the political fight, gathering further momentum, and seeking to reach tipping points where these beginnings of the end start to gain enough traction to become a genuine political project, and not just another harmless daydream or well-intended, soon to be forgotten empty gesture.

As of now the TPNW is a treaty text that courteously mandates the end of nuclearism, but to convert this text into an effective regime of control will require the kind of deep commitments, sacrifices, and perseverance that eventually achieved the impossible, recalling the movements that ending such entrenched evils as slavery, apartheid, and colonialism, but only after long struggles.

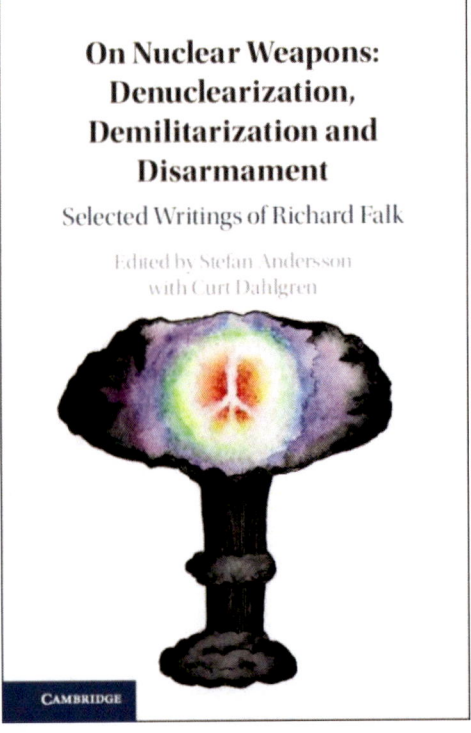

# Rely on science

*José Bustani*

*Ambassador Bustani was the first Director General of the OPCW, who actively sought to bring more countries under the remit of the Chemical Weapons Convention. For his efforts, in 2002, he was illegally removed from office, prior to the US invasion of Iraq the following year (see Spokesman 75).*

*In October 2020, Ambassador Bustani was prevented from testifying directly to the UN Security Coucil about the investigation of the Organisation for the Prohibition of Chemical Weapons into alleged use of chemical weapons in Douma, Syria in 2019. This is what he would have said.*

My name is José Bustani. I am honoured to have been invited to present a statement for this meeting of the UN Security Council to discuss the Syrian chemical dossier and the Organisation for the Prohibition of Chemical Weapons. As the OPCW's first Director General, a position I held from 1997 to 2002, I naturally retain a keen interest in the evolution and fortunes of the Organisation. I have been particularly interested in recent developments regarding the Organisation's work in Syria.

For those of you who are not aware, I was removed from office following a US-orchestrated campaign in 2002 for, ironically, trying to uphold the Chemical Weapons Convention. My removal was subsequently ruled to be illegal by the International Labour Organisation's Administrative Tribunal, but despite this unpleasant experience the OPCW remains close to my heart. It is a special Organisation with an important mandate. I accepted the position of Director General precisely because the Chemical Weapons Convention was non-discriminatory. I took immense pride in the independence, impartiality, and professionalism of its inspectors and wider staff in implementing the Chemical Weapons Convention. No State Party was to be considered above the rest and the hallmark of the Organisation's work was the even-handedness with which all Member States were treated regardless of size, political might, or economic clout.

Although no longer at the helm by this time, I felt great joy when the OPCW was awarded the Nobel Peace Prize in 2013 "for its extensive efforts to eliminate chemical weapons". It was a mandate towards which I and countless other former staff members had worked tirelessly. In the nascent years of the OPCW, we faced a number of challenges, but we overcame them to earn the Organisation a well-deserved reputation for effectiveness and efficiency, not to mention autonomy, impartiality, and a refusal to be politicised. The ILO decision on my removal was an official and public reassertion of the importance of these principles.

More recently, the OPCW's investigations of alleged uses of chemical weapons have no doubt created even greater challenges for the Organisation. It was precisely for this kind of eventuality that we had developed operating procedures, analytical methods, as well as extensive training programmes, in strict accordance with the provisions of the Chemical Weapons Convention. Allegations of the actual use of chemical weapons were a prospect for which we hoped our preparations would never be required. Unfortunately, they were, and today allegations of chemical weapons use are a sad reality.

It is against this backdrop that serious questions are now being raised over whether the independence, impartiality, and professionalism of some of the Organisation's work is being severely compromised, possibly under pressure from some Member States. Of particular concern are the circumstances surrounding the OPCW's investigation of the alleged chemical attack in Douma, Syria, on 7 April 2018. These concerns are emanating from the very heart of the Organisation, from the very scientists and engineers involved in the Douma investigation.

In October 2019 I was invited by the Courage Foundation, an international organisation that 'supports those who risk life or liberty to make significant contributions to the historical record', to participate in a panel along with a number of eminent international figures from the fields of international law, disarmament, military operations, medicine, and intelligence. The panel was convened to hear the concerns of an OPCW official over the conduct of the Organisation's investigation into the Douma incident.

The expert provided compelling and documentary evidence of highly questionable, and potentially fraudulent conduct in the investigative process. In a joint public statement, the Panel was, and I quote, 'unanimous in expressing [its] alarm over unacceptable practices in the investigation of the alleged chemical attack in Douma'. The Panel further called on the

OPCW, 'to permit all inspectors who took part in the Douma investigation to come forward and report their differing observations in an appropriate forum of the States Parties to the Chemical Weapons Convention, in fulfilment of the spirit of the Convention.' I was personally so disturbed by the testimony and evidence presented to the Panel, that I was compelled to make a public statement. I quote: "I have always expected the OPCW to be a true paradigm of multilateralism. My hope is that the concerns expressed publicly by the Panel, in its joint consensus statement, will catalyse a process by which the Organisation can be resurrected to become the independent and non-discriminatory body it used to be."

The call for greater transparency from the OPCW further intensified in November 2019 when an open letter of support for the Courage Foundation declaration was sent to Permanent Representatives to the OPCW to, 'ask for [their] support in taking action at the forthcoming Conference of States Parties aimed at restoring the integrity of the OPCW and regaining public trust'.

The signatories of this petition included such eminent figures as Noam Chomsky, Emeritus Professor at MIT; Marcello Ferrada de Noli, Chair of the Swedish Doctors for Human Rights; Coleen Rowley, whistle-blower and a 2002 *Time* Magazine Person of the Year; Hans von Sponeck, former UN Assistant Secretary-General; and Film Director Oliver Stone, to mention a few.

Almost one year later, the OPCW has still not responded to these requests, nor to the ever-growing controversy surrounding the Douma investigation. Rather, it has hidden behind an impenetrable wall of silence and opacity, making any meaningful dialogue impossible. On the one occasion when it did address the inspectors' concerns in public, it was only to accuse them of breaching confidentiality. Of course, Inspectors – and indeed all OPCW staff members – have responsibilities to respect confidentiality rules. But the OPCW has the primary responsibility – to faithfully ensure the implementation of the provisions of the Chemical Weapons Convention (Article VIII, para 1).

The work of the Organisation must be transparent, for without transparency there is no trust. And trust is what binds the OPCW together. If Member States do not have trust in the fairness and objectivity of the work of the OPCW, then its effectiveness as a global watchdog for chemical weapons is severely compromised.

And transparency and confidentiality are not mutually exclusive. But confidentiality cannot be invoked as a smoke screen for irregular behaviour. The Organisation needs to restore the public trust it once had and which no one denies is now waning. Which is why we are here today.

It would be inappropriate for me to advise on, or even to suggest how the OPCW should go about regaining public trust. Still, as someone who has experienced both rewarding and tumultuous times with the OPCW, I would like to make a personal plea to you, Mr Fernando Arias, as Director General of the OPCW. The inspectors are among the Organisation's most valuable assets. As scientists and engineers, their specialist knowledge and inputs are essential for good decision making. Most importantly, their views are untainted by politics or national interests. They only rely on the science. The inspectors in the Douma investigation have a simple request – that they be given the opportunity to meet with you to express their concerns to you in person, in a manner that is both transparent and accountable.

This is surely the minimum that they can expect. At great risk to themselves, they have dared to speak out against possible irregular behaviour in your Organisation, and it is without doubt in your, in the Organisation's, and in the world's interest that you hear them out. The Convention itself showed great foresight in allowing inspectors to offer differing observations, even in investigations of alleged uses of chemical weapons (paras 62 and 66 of Part II, Ver. Annex). This right is, and I quote, 'a constitutive element supporting the independence and objectivity of inspections'. This language comes from Ralf Trapp and Walter Krutzsch's "A commentary on Verification Practice under the CWC", published by the OPCW itself during my time as DG.

Regardless of whether or not there is substance to the concerns raised about the OPCW's behaviour in the Douma investigation, hearing what your own inspectors have to say would be an important first step in mending the Organisation's damaged reputation. The dissenting inspectors are not claiming to be right, but they do want to be given a fair hearing. As one Director General to another, I respectfully request that you grant them this opportunity. If the OPCW is confident in the robustness of its scientific work on Douma and in the integrity of the investigation, then it has little to fear in hearing out its inspectors. If, however, the claims of evidence suppression, selective use of data, and exclusion of key investigators, among other allegations, are not unfounded, then it is even more imperative that the issue be dealt with openly and urgently.

This Organisation has already achieved greatness. If it has slipped, it nonetheless still has the opportunity to repair itself, and to grow to become even greater. The world needs a credible chemical weapons watchdog. We had one, and I am confident, Mr Arias, that you will see to it that we have one again.

I'm in the process of applying for my Grandmother's Settled Status. She is 84, from Belgium but has lived in Scotland since the 1950s.

But she is so confused with all this and has no idea what is happening. Her two children voted to leave (the EU) and have a cheek to be confused with all this and refuse to explain to her what is going on. I've been fighting with them on every turn with this and finally her application was sent off yesterday, so fingers crossed.

Her whole life is here, she has cried to me how she will never go back to Belgium out of fear of not being allowed back in the UK. It's so hard not to scream at family members who voted for this. But that's pointless now.

She doesn't have a chipped passport so that made things more complicated. I'm really hoping her application is accepted.

**T.M., UK**

Fifteen years of domestic violence have left me destroyed from the inside out. I don't have the energy to fight. I feel completely powerless and insignificant and that triggers the PTSD.

I think the worst thing Brexit did for me was to reinforce that yet again someone else was in charge of my life. My ex used to threaten to have me deported and that I'd never see the kids again… so all that opened the floodgates.

The government's narrative is repeating what my ex told me over 15 years day in and day out… The Home Office with their "your activity is not worthwhile or genuine. Prepare to leave immediately" pretty much describes how I feel about myself after years and years of being told I'm worthless. Just writing these two things makes me cry and I automatically go into fight or flight mode...

**Natasja, The Netherlands**

★★★

I am Bulgarian and my husband is British. Our 12 year old son was born in UK and has dual nationality. I've been living in UK for over 13 years and have always worked. I've studied for five years to get a few diplomas whilst working full-time and being a full-time mum. It was exhausting but I've done it and now have a great job. I never claimed a penny of benefits and payed high taxes. Yet after all I've done, I now get asked to apply for Settled Status like I came here yesterday, even though I was issued a Residence Card by the Home Office back in 2007 to work and live in UK for unlimited time.

It is the same for my mum who's been in UK for 12 years and has the same Residence Card. I've now received Settled Status, but my mum is still getting continuous emails back from the government asking for more and more evidence because they can't track on the system that she's been in the country in the past six months too. But we sent proof! We feel totally messed around by the government. I would like to know why the Residence Cards are no longer evidence.

On top of that, a few months ago I found out that all of my husband's family voted Leave in the 2016 referendum and for Mr Johnson in the latest GE despite the fact that their son, brother, etc. is married to an EU citizen. They all keep saying to me "don't worry you and your mum will be fine, you've lived here long enough"... Well it looks like this doesn't count any more, does it?

I'd like to know if the government realises how many families they are tearing apart as sadly I don't even feel that the relationship with my husband is going in a positive direction now. In the past 3 years I've been feeling awful. I don't know what to do, which path to take. Despite the fact we've built our lives here, that we've created families and our home is here, many want us out.

My mum and I have experienced a lot of xenophobic and racist abuse, and been regularly asked when we are going back to our country. Even in the supermarket we got told by a female customer that it is absolutely forbidden to speak in any other language than English.

**Iva Augarde, Bulgaria**

★★★

I am one of the many who were given "pre-Settled Status" as "not enough evidence can be found" about me on the government's system. I wanted to say that unlike many I didn't accept this and fought it. I sent them more evidence (council tax bills and student enrolment letters) and have received full Settled Status today. Not sure how I feel. I thought I was going to be relieved... but I feel empty and angry instead.

I had already acquired Settled Status! It is an automatic right and I shouldn't have had to go through hell to prove it. I have mainly done it for my two British sons, 5 and 0 years old, in case they settle in the UK, so that their mum can be near them without having to apply for a visa. This whole thing is ridiculous and frankly, quite scary.

**Maria Martinez, Spain**

★★★

I have been struggling to get Settled Status or ILR for my father. He is in a care home now although I looked after him for ten years. His Parkinson's became too much for me to manage on my own as I have significant health problems myself.

I have spoken to the care home about getting him Settled Status but I don't know what is happening. I am struggling to get a conversation with any management. Every time I phone, I am told they are in meetings. I've written letters but have no idea if they have received them.

I know it sounds silly but I have been having nightmares of my Dad being carried out of the care home as he doesn't walk and being forced on to a place.

I don't know what to do.

He should have indefinite leave to remain, which he was told he had when he married my Mum in 1962 but I know things go wrong and information gets lost all the time.

I am so worried.

**Anonymous, Europe**

# Human Skill

## Mike Cooley
## 1934-2020

*John Palmer*

*Mike Cooley*

*John Palmer was Public Affairs Director of the Greater London Enterprise Board and is a former European Editor of* The Guardian. *This is his foreword to* The Search for Alternatives: A Mike Cooley Reader, *published by Spokesman in 2020.*

Prophets and prophecy have not always had a good press. Too often prophetic visions have had limited relevance for those in society and especially the world of work confronting real and immediate challenges. Prophets with their feet placed firmly on the ground and, specifically, those concerned with the so-called 'mundane' world of work have been very rare indeed.

Mike Cooley has every reason to be counted as one such 'prophet.' He has rightly been described by the President of Ireland, Michael D. Higgins, as "… the most intelligent Irish man, the most morally engaged scientist and technologist Ireland has sent abroad." He is certainly a visionary of a future where human skill and labour work in partnership with science and technology rather than in servitude to them.

Born in Tuam, County Galway in Ireland, Mike Cooley studied advanced computer based engineering in Germany and Switzerland. He first came to public attention as a result of his pioneering role in the British trade union movement as an advocate of human skill being enhanced by and not harnessed or displaced by technology.

As a trade unionist working in the Lucas Aerospace company he played a key role in outlining how workers could confront the threat of mass redundancies by showing how their skills could be adapted to produce alternative "socially useful" products and demonstrated practical examples in health, transport and other sectors.

The socially responsive ethos of the human-centred movement generated by the Lucas Workers' Plan of 1976 is summed up in the Mike's statement that "there cannot be islands of social responsibility in a sea of depravity". He also warned about "the appalling gap between what technology could provide for society, and what it actually does provide."

> "The tragic waste our society makes of its most precious asset—the skills, ingenuity, energy, creativity and enthusiasm of ordinary people"; and "the myth that computerisation, automation and use of robotic devices will automatically free human being from soul destroying, backbreaking tasks and leave them free to engage in more creative work."

Mike Cooley's vision was of a human-machine symbiosis as an alternative potential for work life. He saw this as part of wider European humanistic movements such as 'Democratic Participation' (Scandinavia) and 'Humanisation of Technology and Work' (Germany). These European human-centred movements provided a basis for the establishment of the 'Anthropocentric Systems and Technology' programme of the European Union during the 1990s.

If his approach had received the political support in the wider labour movement it deserved, perhaps the worst depredations of Thatcherite 'slash and burn' economics in the 1970s and 80s might have been mitigated or avoided. In later years his role as Director of Technology in the innovative Greater London Enterprise Board (GLEB) allowed these ideas to reach a wider audience.

But its work was undermined when Margaret Thatcher's government closed the Greater London Council which had sponsored the creation of the GLEB. Alas the longer term benefits of the radical strategies Cooley and others canvassed were more often realised in other European countries rather than the UK. It was left to others to invest in projects such as a pioneering 'road/rail bus' and a new type of portable kidney machine. We are still paying the price today in the post-industrial desert created by the 'free' market, particularly in the former industrial areas of Britain. The economic deprivation and social inequality generated by the free market system has spawned new forms of extreme right populism.

Mike Cooley has also played a crucial role in developing thinking about how the interplay between the diversity of human skill and the calculation capacity of the machine can lead to enhanced productivity and enriched human expertise, combining human ingenuity and technological innovations.

Cooley has warned us of the danger of the objectification of human knowledge and experience into information and data, risking human judgement becoming mere calculation and turning the human into a mere robot. This was dramatically expressed in his pioneering book *Architect or Bee?* Alas neo-liberal capitalism has taken us down a very different path. The monumental squandering of the creative potential of working people – in partnership with human centred science and technology – has led to the emergence of casualised labour in both industries and services. Far from productivity being given a massive boost, too many people work in low productivity sectors, low paid and often without adequate protection from unemployment and fluctuating income.

When Mike talked about the complex and little understood relationship between a worker's innate skills and the tools and technologies available to them, he would draw on a rich knowledge of history. "Think of the breathtaking achievements of the mediaeval workers who built the great cathedrals. Who were the architects?" we heard him ask.

His response was clear:

"Actually there was no separate cast of architects giving instructions to a passive work force. Rather every stonemason and building worker had an innate sense of the potential design born from long and intimate knowledge of the materials with which they were working."

The impressive scope of Cooley's thinking is well reflected in his subsequent books *The Human Price of Technology*, republished in 2016, and *Delinquent Genius: The Strange Affair of Man and His Technology* which was published in 2018 by Spokesman. He was quick to identify the gender bias in the pattern of contemporary work and skills distribution.

The thrust of Mike Cooley's human society focused analysis has a striking parallel in the rapidly growing world-wide movement – led by young people – against climate change and for radical, green policies in all the major aspects of our economic, social and individual lives to counteract it. It is a development profoundly welcomed by him as the climate change threat to our planet and its people looms ever larger. In a sustainable world economy, the values and goals of Michael Cooley's work on human centred technology are sure to be reflected.

# The Shout of Joy

*Mike Cooley*

*From* Delinquent Genius: The Strand Affair of Man and His Technology, *published by Spokesman in 2018.*

In the narrow, rationalistic, technological, programmable world I have been talking about, we are continually required to deny the reality of our own experiences. This takes the form of a post hoc rationalisation of what we have experienced. In order for it to be significant, we are required to show that it is the outcome of a series of logical sequential steps which led to an end result. I am not here suggesting that such logical sequential steps do not have their part to play nor am I suggesting they are insignificant. I am however suggesting that they are but a tiny part of the totality of human experience. However, even in research and development laboratories where large groups of teams are working together, innovations, concepts for new ideas are still based on the insights of individuals and that those insights and ideas are the stimulus for a research and development activity or for the identification of a product which may ultimately result.

Since, however, insight like imagination and intuition is not quantifiable, is not predictable, and is not repeatable, its significance is always underplayed if not totally denied. This denial is built in at the very root of our scientific methodology, a methodology which is dominating our thinking in all fields of human endeavour. We all of us tend to construct an edifice of post hoc scientific explanation since deep down we have a fear of that uncertainty and that unpredictability which is at the root of intuition, imagination and creativity. Let us suppose for a moment that we encounter a great and acclaimed scientific thinker who has codified the laws of thought and in doing so has laid the basis for the use of mathematics as a model of thinking. Let us suppose that this great scientific thinker has produced a philosophical and scientific framework for rationality. Let us suppose that following profound scientific and philosophical discussion, these ideas were found to be internally consistent within a logical framework. Let us further suppose that this framework was the basis for mathematical modelling of thought processes which showed that thinking could be modelled and these models could be based on mathematics. Let us then suppose that this work had been massively funded worldwide to the level of billions of dollars annually and all of this work demonstrated in its own self consistent terms that there were logical, scientific and mathematical ways of modelling thought processes. Suppose then we were to meet the originator of this huge and mathematical scientifical edifice of analysis and we asked them how they arrived at their concept of this scientific construct. We would be alarmed if they told us "I was told by an angel". Our alarm would not be lessened if they added that they were told by an angel in a dream, yet it would seem that that was exactly what did happen. On the night of November 10th 1619 Daca had a series of three dreams in which the angel of truth revealed to him a secret which "would lay the basis of a new method of understanding and a new science".

The significance of such dreams, revelations, brainwaves, insights, sparks of imagination are usually carefully omitted from post hoc analysis. It seems to be slightly more acceptable to refer to them in the context of the arts. To do so is indeed of longstanding. Socretes pointed out: "The authors of these great poems which we admire do not attain to excellence, to the rules of any art. They utter their beautiful melodies of verse in a state of inspiration as it were possessed by a spirit of their own". (*The unknown guest* p. 7)

Stephen Spender, whilst emphasising the importance of "Work" on a poem, states: "Inspiration is the beginning of a poem and it is also its final goal ... My own experience is certainly that of a line or a phase or a word or something still vague, a dim cloud of an idea which I feel must be condensed into a shower of words". He points out that Paul Valéry speaks of the "una ligna donne/e" of a poem. One line is given to the poet by God or by nature. The rest he has to discover for himself. The significance of childhood and early images is emphasised when Spender says: "All poets have this highly developed apparatus of memory and they are usually aware of experiences which happened to them at the earliest ages and which retain their pristine significance throughout their lives." One might refer to Dante's meeting with Beatrice which became the symbol around which the devine comedy crystallised. The experiences of nature which form the subjects of Wordsworth's poetry were an extension of a childhood vision of natural presences which surrounded the boy Wordsworth. The decision in his later life to live in the Lake District was one based on the desire to live on those childhood memories which were the most important experiences in his poetry. One might cite Rilke's notion of words rising up and imposing themselves upon him and they being "the most mysterious". One might cite Mozart's famous letter, although some tend to discount it or Tchaikovsky for whom music is "a more subtle medium in which to translate the 1000 shifting moments in the mood of a soul". One might fill many moments with examples of the importance of inspiration, mood, insight, imagination and that inner creative voice which so often sparks our imagination in the most unlikely places and at the most extraordinary times. A professor of anatomy at Harvard summed it up succinctly over 100 years ago when he said: "We all of us have a double who is wiser and better than we are and who puts thoughts into our heads and words into our mouth". This double he describes as:

> "A creating and informing spirit which is with us and not of us, is recognised everywhere here and in story life. It is the Zeus that kindled the rage of Achilles; it is the muse of Homer – it is the demon of Socrates, it is the inspiration of the see-er – it is the smoking devil that whispers to Margaret as she kneels at the altar – it is the hobgoblin who cries 'Sell him, Sell him' in the ear of John Bunyan; it shaped the forms that formed the soul of Michelangelo when he saw the figure of the great law giver in the yet unhewn marble in the dome of the world's yet unbuilt Basilika against the black horizon. It comes to the least of us as a voice that will be heard; it tells us that we must believe; it frames our sentences; it lends a sudden gleam of elegance to the dummest of us all."

It may be accepted that this inspiration and intuition is significant in the arts but has no place within the sciences, indeed with the technocratic reductionist thinking which now dominates Western society. Little is done to provide an enviroment which stimulates the imagination and leads to creativity in its multivarious forms. Given the austere, logical, sequential, mathematical image of science, it is worth emphasising that creative scientists, with few exceptions, freely admit to the significance of intuition, insight, and imagination.

In the world of the reductionist, everything has to be made clear, specific and precise. There is no space for ambiguity, uncertainty or lack of clarity. The history of ideas suggests it is not as straightforward as that. Frequently it is some half-baked idea, some fuzzy correlation, a smell, a sound, that triggers the imagination and produces something of significance.

In the reductionist view, everything can be explained by scientific means. This means that every aspect of human behaviour must be explored and expressed scientifically. Our innermost thoughts, our likes, our dislikes, should all be made visible and explained.

Even our dreams will now be subjected to such scrutiny. Research is now taking place into what is known as 'lucid dreaming'. The object, as with most science and technology, is to achieve control. In a lucid dream you become conscious that you are dreaming and you take control. There has been developed the dreamlight, which it is held, will help people to have lucid dreams. In the 50s, psychologists rejected that notion, when electroencephalogram revealed the faces of sleep, especially REM (rapid eye movement) sleep, during which, if you wake somebody they are likely to report that they have been dreaming. The muscles are relaxed to the point of paralysis, the brain is active but the sleeper is difficult to wake.

It was held that it would be impossible to be conscious during such a dreaming state. Development suggests that by using REM techniques, it is possible to know when one is dreaming and hence make estimates of the time of the dream.

It is said that this research will give insight into the ultimate human mystery, the brain, by identifying the range and location of its activities. It is believed that lucid dreaming will act as a bridge between the conscious and the sub-conscious, and that the work could have surgical as well as psychological benefits. It is suggested that the same parts of the brain are active in waking and sleeping. Tasks such as singing use the opposite half of the brain. Researchers have found that subjects signal from a dream that they are singing when in fact they are adding up. The same areas are active.

With the dreamlight, the eye movements are detected using personal computers which switch on flashing lights when the eye movements rapidly reach a certain point. It is hoped that eventually all the necessary circuitry will be incorporated in a pair of goggles. These are put on at bedtime and flash only when the subject is asleep and dreaming. The person can control the level of eye movements at which the light begins to flash.

Francis Crick, who won the Nobel Prize for his shared discovery of DNA, suggests there is a danger in this interfering with dreams. He believes that the function of dreams is forgetting, in which case it might be best that they are left alone. However, researchers working on the dreamlight are adamant that "its going to help people have lucid dreams and lucid dreams are good" (*Sunday Times*, June 6th, 1990).

In considering the work of scientists, engineers and mathematicians, it may seem sacrilegious to talk about something as introverted and personal as their "spirit", yet scientists themselves would go even further. Poincaré suggests it is necessary to "see what goes on in the very soul of the mathematician". To illustrate the point, he described how he discovered a number of mathematical theorems:

> "For 15 days I strove to prove that there could not be any functions like those I have since called Fuchsian functions. I was then very ignorant. Every day I seated myself at my worktable, stayed an hour or two, tried a great number of combinations and reached no result. One evening, contrary to custom, I drank black coffee and could not sleep. Ideas rose in crowds. I felt them collide until pairs interlocked so to speak, making a stable combination. The next morning I had established the existence of a class of Fuchsian functions – those which come from the hypergeometric series. I had only to write out the results which took me but a few hours."

In extending this area of work, he likewise had further 'insights':

> "The changes of travel made me forget my mathematical work. Having reached Coutances, we entered an omnibus to go to some place or another. At the moment I put my foot on the step the idea came to me without anything in my former thoughts seeming to have paved the way for it, that the transformations I had used to define the Fuchsian functions were identical with those of non-Euclidian geometry."

And without any notion that they may be connected to his preceding researches, he states:

> "Disgusted with my failure. I spent a few days at the seaside and thought of something else. One morning walking on the bluff, the idea came to me with just the same idea of brevity, suddenness and immediate certainty, that the arithmetic transformations of intermediate, ternary quadratic forms were identical to those of non-Euclidian geometry."

I am not suggesting here that such sudden insights will come to just anybody. Clearly a deep knowledge of mathematics and the study of those subjects is a precondition for the ability to recognise such an insight when it occurs. The point I am making here, however, is that the mere possession of mathematical knowledge acquired in a systematic rule based way, is not the whole story.

Furthermore, such rule based systems do not take into account that which goes on in the subconscious. Poincaré points out: "Most striking at first is this appearance of sudden illumination, a manifestation of long unconscious prior work. The role of this unconscious work in mathematical invention appears to me incontestable, and traces of it can be found in other cases where it is less evident".

The point I wish to emphasise here is that he repeatedly points out the importance of the emotional and the feeling in the context of mathematical creativity, and that which he called privileged unconscious phenomena:

> "Those susceptible to becoming conscious are those which directly or indirectly, affect most profoundly our emotional sensibility ... It may be surprising to see emotional sensibility evoked apropos of mathematical demonstrations which it would seem can interest only the intellect. This would be to forget the feeling of mathematical beauty, of the harmony of numbers and forms, of geometric elegance. This is a true aesthetic feeling that all real mathematicians know and surely it belongs to emotional sensibility."

Computerised systems, and those which are rule-based by definition, lack this true "aesthetic feeling" and "emotional sensibility". Thus in spite of the speed, consistency and repeatability, we should not think of them in the sense of human progress and creativity, as superior to human beings but rather they should be treated as support systems and tools which support the creative emotions of human beings. Sadly, most scientists seem unwilling to admit to the significance of this emotional and subjective dimension of their work. It is perhaps a measure of the extent to which they are themselves victims of the dominant technocratic ideology. It is also surprising that a profession which prides itself on its rationality and objectivity so often seeks to deny the truth of the creative process.

They do seem to be victims of their own propoganda and determined to imply that scientific advance, the emergence of mathematical theorems and the invention of new products are based on a scientific method which involves inductive reasoning, logical sequential steps, in a word something which is rigidly codified and leaves little space for our humanity. This in turn gives rise to the notion that scientific "training" at school and university should be based on hard facts, methods, procedures and that excitement and motivation have little place in the preparation of scientists. It means that that part of us which is the essence of our humanity has little space within the sciences and that creative people should find vent for their capabilities elsewhere.

In reality, science should be an exciting as well as a demanding area of study which has space for both the subjective and the objective, the inductive and the deductive. Scientists actually succeed in alienating potential students of their courses with the image they create of their profession. Medua points out that if one really presses a scientist, they would "Probably mumble something about induction" and "establishing the laws of nature"; but if anyone working in a laboratory professed to be trying to establish the laws of nature by induction we should begin to think that he was overdue for leave. It is most unlikely that anything more than a tiny minority of theorems were ever arrived at, "discovered" merely by the existence of deductive reasoning. Most of them entered the mind by processes of the kind vaguely called intuition. Deduction or logical derivation came later, to justify or falsify what was in the first place an inspiration or an intuitive belief. This is seldom apparent from mathematical writings because mathematicians take pains to ensure that it should not be.

I would go so far as to say there is an intellectual dishonesty amongst such professions. If they wish to be dishonest with each other, that is up to them. What is unacceptable is that their dishonesty is used as part of a wider process which is increasingly imposing on society this narrow mechanistic view of the world. Much more needs to be done by scientists themselves to have the courage to allow their inner voices to speak. On the occasions when they do, we get some of that human freshness which is almost poetic. The discoverer of insulin, Sir Frederick Banting, says that ideas come "when the imagination is allowed to run riot on the problems that block the progress of research, when the huge stones of scientific fact are turned over and fitted in so that the mosaic figure of truth, designed by mother nature long ago, may be formed from the chaos".

It is important that we freely admit to Ampere's "shout of joy" when he finally found the solution to a problem he had firstly formulated some seven years earlier. Our science and technology would be the richer, and our courses in those subjects would attract the more creative, sensitive members of society, if they included accounts of Faraday having a vision of tubes which "rose up before him like things" or admit with Carl Gauss:

> " ... come to me when my mind was fatigued or when I was at my working table. They came particularly readily during the slow ascent of wooded hills on a sunny day."

In all of these cases, it was attributes other than just technical and scientific competence that appeared to have facilitated the 'Eureka' step. Certainly, there was a sensitivity to a problem or an issue. There is usually much fringe consiousness of the subject area but above all there is imagination and it is this which in many ways defines our humanity. Gaining the power to accumulate experience and to reason was not enough to make him man. Another quality was necessary – the great gift of imagination. "This is perhaps man's most distinctive trait for it makes possible his creativity." The dominant tendancy in our society is to deny the significance of the subjective, the imaginative and seek to replace that by the scientific and the quantifiable rather than to complemenmt it by this rational dimension.

I do believe the long term implications for our species are serious indeed. We have associated the ascent of man with imagination and creativity. It seems not unreasonable to assume that we can associate the decline of man with the lack of these attributes.

# The Search for Alternatives

*Liberating Human Imagination*

## A Mike Cooley Reader

"The thrust of Mike Cooley's human society focused analysis has a striking parallel in the rapidly growing worldwide movement – led by young people – against climate change and for radical, green policies in all the major aspects of our economic, social and individual lives to counteract it. It is a development profoundly welcomed by him as the climate change threat to our planet and its people looms ever larger. In a sustainable world economy, the values and goals of Michael Cooley's work on human centred technology are sure to be reflected."

John Palmer

£11.99 | 202 Pages | ISBN: 978 085124 8851

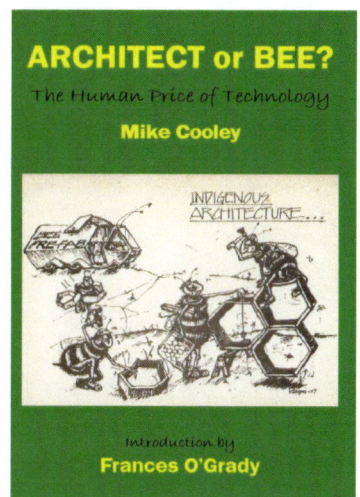

Price: £10.99 | 194 pages
ISBN: 978 0 85124 8493

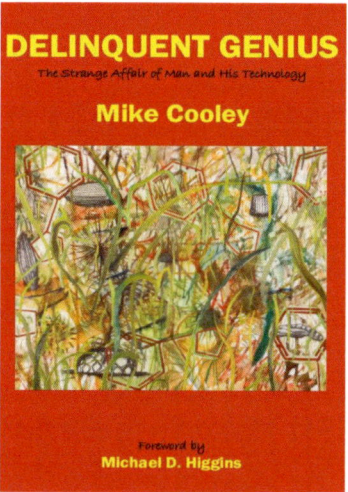

Price: £11.99 | 248 Pages
ISBN: 978 085124 8783

# www.spokesmanbooks.com

# The Financialisation, Marketisation & Privatisation of Renewable Energy

**626** transactions in global renewable energy secondary market in 20 months cost **US$289bn**

## Strategies for public ownership

Dexter Whitfield
ESSU Research Report No. 12

# Summary of findings

The research identified 626 renewable energy secondary market transactions in the global economy in a 20-month period between January 2019 and 31st August 2020 (326 transactions in 2019 and 300 in the first 8 months of 2020). They primarily reflect the sale of wind parks, solar farms, hydro, biomass, energy-from-waste and battery storage projects that are at the development, construction or operational stage.

Table 1: **Global overview of renewable energy transactions in 20 months**

| Key Findings | |
| --- | --- |
| Total transactions | 626 |
| Cost | US$289bn |
| Total megawatts | 300,000 |
| % private equity | 34% |
| % public sector | 4% of MW or 4.4% of transactions |

Source: ESSU Global Renewable Energy Secondary Market Transactions Database, 2020.

If the current rate of transactions increases 10% annually in the next ten years and assuming an average 1% increase in inflation, there could be nearly 10,000 transactions at a cost of US$4,825bn.

The scale of transactions is a direct reflection of the extent of private sector ownership and operational control of renewable energy projects. Governments and public sector companies accounted for only 4.4% of these transactions which reinforces the evidence that the ownership and operation of the renewable energy sector is dominated by the private sector.

Governments and public authorities are involved in renewable energy policy making, auctions of project sites, approving planning permission, subsidies and guarantees intended to attract private investment. The World Bank, the International Finance Corporation, regional development banks and overseas aid agencies provide direct finance or loans to similarly attract private investment in the global south. It is a repeat of the corporate welfare role which became common in other sectors over the last 40 years.

Other key findings include:

- Transactions were concentrated in Europe and North America which account for 315 and 179 transactions respectively and virtually the same level of 98,000MW but includes transactions in Asia, Latin America, Africa and Oceania.
- Private equity funds were involved in 159 transactions acquiring renewable energy assets and 53 transactions in selling assets.
- Utility companies were involved in 44 transactions where they purchased renewable energy assets and 18 where they sold assets.
- Petroleum companies were involved in 16 acquisitions and 7 transactions where they sold assets.
- 140 transactions involved a parent company or subsidiaries registered in tax havens.
- 38 assets were acquired into public ownership (15,270MW), 14 assets were privatised (14,504MW) and 14 assets were transferred between public authorities (5,175MW) – there was a net gain in the number assets transferred to the public sector there was only a small gain in the level of MW.
- There was a wide variation in type of projects (wind, solar, hydro, biomass, energy from waste, battery storage) size (1MW - 10,000MW) and cost (€1.1m – US$4,668m) of transactions.

## Increase publicly owned and operated renewable energy generation

1. Prioritise publicly financed renewable energy generation in industrialised countries and with a wider generation, grid and supply in developing countries.
2. Increase direct public investment in new renewable energy projects and retain them in public ownership and operation.
3. Agree selective public acquisition of renewable energy assets by negotiation or via nationalisation. All future public financial support must be conditional on binding agreements that give the public sector the first option to acquire the project in any future sale of the project.
4. Increase public sector in-house capabilities to plan, develop and operate renewable energy projects. A decommodification process (redesign of services, jobs, regulations, democratic accountability, participation and disclosure) in government and public authorities should be combined with the adoption of radical public management committed to public ownership and provision. This would terminate the financialisation, marketisation, individualisation and privatisation processes and significantly improve the quality of public services and terms and conditions and training of public employees (Whitfield, 2020a).
5. Ensure regeneration and development plans, strategies and retrofitting housing programmes include new publicly owned and operated renewable energy projects.
6. Classify renewable energy assets as public goods and align with public service principles and values.
7. Maximise local and regional economic development and employment opportunities created by the construction, production and operation of renewable energy projects and make quantified proposals a condition of regulatory approval and any direct or indirect public financial support.
8. Ensure that innovative and technological developments are harnessed to meet collective social, economic, environmental and power generation needs instead of driving market forces and profit maximisation.
9. Demand more stringent policies that ban the use of tax havens for the finance and ownership of renewable energy projects.
10. It is imperative that Green Deals contain detailed proposals to address the operation and impact of the secondary market in renewable energy projects within nation states and internationally.

These proposals are set out in Part 3 with strategies to widen the scope and powers in regulatory frameworks to increase democratic accountability, obtain greater local/regional economic and employment benefits from renewable energy investments.

It proposes the establishment of a National Renewable Energy Agency to strengthen economic linkages between projects and local/regional economies, coordinate research and innovation and to address training and skills development.

Download the full report at the European Services Strategy Unite website:
www.european-services-strategy.org.uk

# Dexter Whitfield titles from Spokesman Books

## Public Alternative to the Privatisation of Life
*Dexter Whitfield*

This book focuses on privatisation, automation and megacities, and the political economy of ten forms of privatisation. It provides detailed global evidence of the high costs and consequences for public services, assets, jobs and inequalities and the business of extracting profits from public assets and service users. The final part of the book details proposals for democratisation and participation together with decommodification, which is essential to enable effective public ownership and provision. Public assets and services under neoliberal public management would be disastrous, hence a new radical approach to public management is essential..

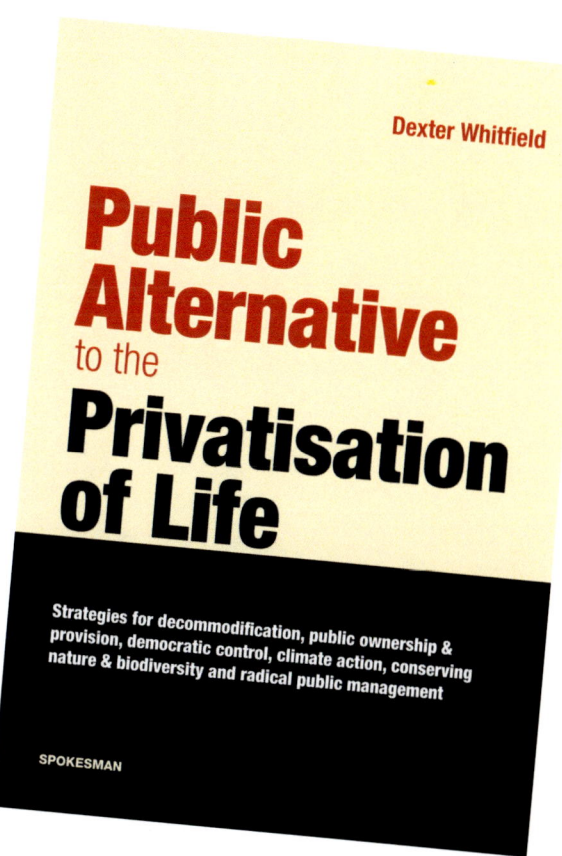

Price: £25.00 | ISBN: 978 0 85124 883 7
580 Pages | Paperback

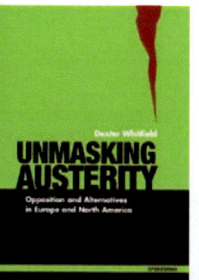  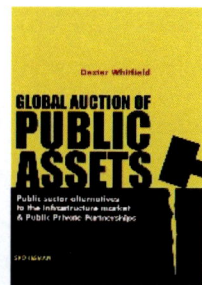

**Unmasking Austerity**
£15 | 128 Pages | A4
**In Place of Austerity**
£18 | 264 Pages | A5
**Global Auction of Public Assets**
£18 | 380 Pages | A5

# www.spokesmanbooks.com

"An unparalleled resource." —NAOMI KLEIN

# ASYLUM FOR SALE

## PROFIT AND PROTEST IN THE MIGRATION INDUSTRY

EDITED BY
**SIOBHÁN McGUIRK**
AND **ADRIENNE PINE**

FOREWORD BY
**SETH M. HOLMES**

# The Business of Selling Life: Reflections from a Rescue Ship in the Mediterranean Sea

## Alva, Uyi, and Madi

**The *Aquarius***

The *Aquarius* was a search and rescue ship that patrolled the Mediterranean Sea from February 2016 to October 2018.[1] It was funded by donations, operated by Médecins Sans Frontières (MSF), and run by volunteers. It takes ships like the *Aquarius* three days to cover the three-hundred-mile stretch of water that separates Libya and Italy. In Libya, smugglers tell people waiting to cross that the journey takes three hours.

A rubber vessel carrying approximately one hundred people, seen from the *Aquarius*. Often, only men are visible at first. Women and children on board usually sit in the middle of the boat, where they are at risk of suffocation if the floor cracks. Corrosive puddles of fuel and salt pool on the floor and burn people's skin.

Most of the vessels intercepted by rescue ships in the Mediterranean are flimsy rubber boats designed for forty people and loaded with over a hundred. They are not intended to withstand the journey to Europe, carrying just enough fuel to sail a few miles from the coastline. They are either intercepted by Libyan authorities and returned to Libya, rescued by a nongovernmental organization (NGO) ship, or they sink.

NGO ships play an important role in the Mediterranean, one that was previously filled by European states. In 2014, the European Union pressured the Italian government to end its Mare Nostrum (Our Sea) program, which had previously provided rescue boats, food, and medical and legal services to migrants rescued at sea. Frontex, the EU's border security agency, replaced Mare Nostrum with Operation Triton and a new mandate: "search and rescue" became "border control." Triton patrols the Italian coast—too far from Libya to reach vessels in distress before they and their passengers sink.

In 2017, the Italian government introduced a code of conduct for NGO-sponsored boats, in an attempt to ban the transfer of refugees to larger rescue ships and to force NGO crews to allow police officers on board.[2]

Such policies have been met with dismay by human rights advocates. The German charity Sea-Watch accused the EU of "willfully letting refugees drown," while Amnesty International characterized the code as a "concerted smear campaign" against NGO rescue operations.[3]

Governmental decisions to defund rescue operations and to "cooperate" with Libyan authorities—all under the rubric of "border control"—have had deadly consequences. In 2016, an estimated five thousand people died attempting to cross the Mediterranean to Europe.[4] Thousands more drowned in boats that sank unseen. People intercepted by the Libyan authorities or deemed by Frontex adjudicators to be "irregular migrants"—the vast majority of those who survive the journey and make it to Italy—are "sent back" to state-run detention centers in Libya,[5] where they face enslavement, captivity, and violent exploitation.[6]

Critics argue that rescue boats incentivize crossings. State governments and EU bodies promote this view, critiquing and intervening in independent search and rescue operations. People making the crossings tell a very different story.

**"You Pay to Die"**

On June 29, 2016, the *Aquarius* intercepted a vessel in distress carrying 111 passengers. Uyi,[7] an artist from Nigeria, was on board. A few hours after the rescue, he drew these images and asked us to photograph them, "so people can see how hard it was making this journey and how we were maltreated."

**Uyi:** This picture shows the start of the journey crossing the border between Nigeria and Niger by motorbike. It was not an easy decision to leave Nigeria, and getting money to embark on the journey was not easy. It took a lot of time and determination. I spent two months on the road from Nigeria to Libya, and it was terrible, horrible. You pay for every step you take. You have to pay. If you don't pay, you cannot go to the next place. You see people dying on the road. I never knew it would be so dangerous to leave Nigeria.

**Uyi:** We reached Agadez in Niger. That's where you meet the people who will take you on to the next stage, the smugglers who will take you to Libya. From Agadez, we took a vehicle to Libya. It took us through the Sahara Desert to Qatrun. There were twenty-five people in the vehicle. Women and children too. You can see how we were sitting. Each of us was sitting holding these wooden sticks. If someone falls off, they keep driving. I was in that car for over five days. No water for days while you are in the desert. I felt so bad and so weak. When I got down [from the vehicle], I couldn't even stand. My legs were shaking. It was a bad experience.

**Uyi:** I was in Libya for close to four months. I was in Sabha, and then in Tripoli, where I did some small jobs. To raise money there was not easy. Then I fell into the hands of other men. They maltreated us. We stayed in a house that was like a prison, where you could not go out; you could not take a shower. I ended up in this house after I met some African guys in Libya. I wanted to make this journey to Europe, and I asked them where they were going. They were also heading this direction, so I thought I would give them the money, and they would bring us here. I didn't know that after giving them the money they would take us to this prison and keep us there for months. They treated us like prisoners and slaves. They were not humans who kept us. They gave us a handful of pasta and saltwater from the sea to drink. They beat people so much. So much. Why? They beat us for talking. They beat you for money. They don't have a reason to beat you. They beat you for no just cause. Because they don't hear you speak English, they beat you. When they wanted us to work, they put us in a van and took us to work somewhere. There were over six hundred people in that place, all different nationalities. Some of those rescued here today [by the *Aquarius*] were in that prison too. From there, they took us to the beach at night and put us in a *lampedusa* [a rubber dinghy, named after the Italian island where many boats land]. You couldn't raise your head to look. If you did, they beat you.

**Uyi:** A lot of thoughts were going through my mind on this boat, like: "If this thing explodes now, I will die. The water will not help me. Oh, God, please send a rescue. Come save me. Oh, why did I put myself in this mess? No land to put my leg on." Look this way, that way—all sea. A lot of thoughts on my mind that day. It was not easy being on that boat. We stayed on that boat for what felt like days. It was so horrible. You pay to die. That is how it is: you pay to die. We prayed for rescue.

**Engendered Dangers**

A common misconception is that most people attempting the crossing are adult men. The UNHCR estimates that nearly half are women or children.[8] Smugglers advise people to say that they are over twenty, because adults are taken to less secure processing centers in Italy—from which it is easier to "disappear." Fifteen-year-old Madi travelled alone from Mali.

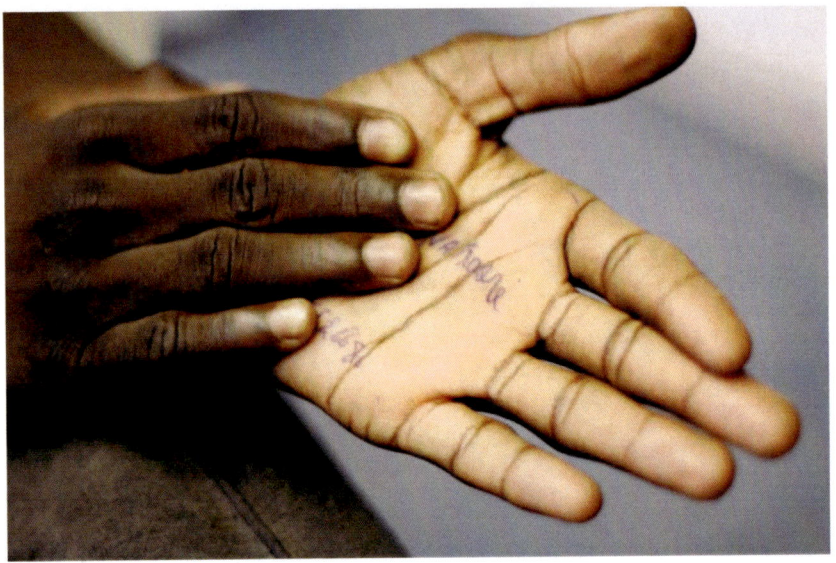

**Madi:** I went to primary school and I worked on a farm. I travelled in a big truck from Mali to Libya, with sixty-one others. First, I went from Mali to Algeria, and then to Libya. I paid money to get to Libya. On the first day of Ramadan, they put me in prison. They beat me a lot. Under my feet and on my body. No good food and very bad treatment. It was like a very large prison with a metal roof over the top and no windows. Just one door and a little bit of light from where the sun could get in on the sides of the roof. Bandits ran the prison, carrying small pistols and big long guns. I saw dead bodies in that prison. They died of starvation or illnesses. [The guards] held me at gunpoint in the prison to give them money. I paid them so that I could escape. My father is in France. He has been there for eight years. I have his phone number written on my hand. I spoke to him when I was in Mali, and he told me not to make this journey. I said I want to come. My father said that it is too dangerous. But I chose to come. He does not know that I made the journey over the sea. How else could I come?

During their journeys, women are exposed to high levels of sexual violence at the hands of smugglers, private individuals, armed groups, militias, criminals, and traffickers. Young Eritrean women told medics on the *Aquarius* that they had been advised to use injectable contraception before starting the journey, as the risk of rape was so high—particularly in the desert and in Libya. Other women did not receive such advice. While on board, they were offered pregnancy tests. Some found out they were pregnant as a result of the rape they had endured reaching Europe. One woman on the ship wanted to take her own life after finding out that she was pregnant.

**Fortress Europe**

After intercepting a boat in distress, the *Aquarius* travelled back to Europe. When the ship pulled close to port, all on board fell quiet. The mood sank from elation at surviving the crossing to uneasy anticipation. A wealth of hopes, dreams, skills, resilience, and youthful vision arrived in the port—met by administration tents and officials ready to "vet" their owners.

In December 2018, MSF and its partner SOS Méditerranée were forced to terminate operation of the *Aquarius*. MSF announced that the decision "was the result of a sustained campaign, spearheaded by the Italian government and backed by other European states, to delegitimize, slander and obstruct aid organizations providing assistance to vulnerable people."[9]

Eritrean women arriving at the port of Catania, Sicily, on August 23, 2016. The *Aquarius* had taken them aboard, along with over four hundred other people who had been crammed into an unseaworthy old wooden fishing boat. This was their first sight of Europe.

As of January 2019, no dedicated rescue boats were operating in the Central Mediterranean. The UNHCR has warned that while the overall number of people attempting the journey was decreasing, the rate of those dying at sea was rising. In 2018, six people a day died trying to cross the Mediterranean.[10] The UNHCR concludes that cuts in rescue operations by European countries and restrictions on humanitarian boats are pushing up the death toll. By reducing documentation and publicity of vessels in distress, as well as rescue operations themselves, Europe can claim "successes on migration," while thousands continue to drown in its waters.[11]

**Alva White** works as a BBC journalist when she is not on assignment as a field communications manager for Médecins Sans Frontières.
**Uyi** is an artist from Nigeria.
**Madi** is a student from Mali.

**Notes**
1. All photos by Alva White.
2. Lizzie Dearden, "EU Accused of 'Willfully Letting Refugees Drown' as NGOs Face Having Rescues Suspended in the Mediterranean," *Independent*, July 29, 2017, accessed March 21, 2020, http://www.independent.co.uk/news/world/

europe/refugee-crisis-ngo-rescue-ships-mediterranean-sea-italy-libya-eu-code-of-conduct-deaths-2300-latest-a7866226.html.

3   Dearden, "EU Accused of 'Willfully Letting Refugees Drown' as NGOs Face Having Rescues Suspended in the Mediterranean."

4   "Desperate Journeys: Refugees and Migrants Entering and Crossing Europe via the Mediterranean and Western Balkans Routes," UNHCR Bureau for Europe, February 2017, accessed March 21, 2020, https://www.unhcr.org/news/updates/2017/2/58b449f54/desperate-journeys-refugees-migrants-entering-crossing-europe-via-mediterranean.html.

5   Alan Travis, "EU Summit to Offer Resettlement to Only 5,000 Refugees," *Guardian*, April 23, 2015, accessed March 21, 2020, https://www.theguardian.com/world/2015/apr/22/most-migrants-crossing-mediterranean-will-be-sent-back-eu-leaders-to-agree.

6   *A Perfect Storm: The Failure of European Policies in the Central Mediterranean*, (London: Amnesty International, 2017), accessed March 21, 2020, https://www.refworld.org/pdfid/597f0fed4.pdf.

7   Both Uyi and Madi requested to use their first names only.

8   "Desperate Journeys."

9   "Aquarius Forced to End Operations as Europe Condemns People to Drown," Médecins Sans Frontières, December 6, 2018, accessed March 21, 2020, https://www.msf.org/aquarius-forced-end-operations-europe-condemns-people-drown.

10  "Desperate Journeys."

11  "Aquarius Forced to End Operations as Europe Condemns People to Drown."

Asylum for Sale, *edited by Siobhán McGuirk and Adrienne Pine, is published by PM Press (pmpress.org.uk). We are grateful for permission to print this extract, which develops our coverage in* Spokesman 134.

# Stephen F. Cohen
## 1938 – 2020

For most of his life Stephen F. Cohen was out of step with the establishment view of Russia, but it was with the rise of President Putin that this divergence became a chasm. The days of the inebriate Boris Nikolayevich Yeltsin were too much for him and too much for Russia: some semblance of order had to be restored. It had been Cohen's job to explain Soviet Russia, Gorbachev's and Yeltsin's Russia, and now Putin's. The latter perhaps could be described in Churchill's remark that Russia was 'a riddle, wrapped in a mystery, inside an enigma', or did the Western media and governments want Putin's Russia to be seen that way — sinister, inherently corrupt, different from the West, and above all expansionist? It fell to Cohen to bring understanding and, when necessary, to counteract the lies and misinterpretations peddled by the anti-Russian propaganda machine.

Cohen's achievements are many, but the late period of his life, when he took on the establishment of the Number One superpower, stands testimony to his personal bravery. For many years he was undoubtedly a thorn in the side of the US government (particularly the State Department) and the academic establishment. With his courteous and gentle manner, armed with a different appraisal of the Bolshevik Revolution and the following regimes, and admiration for the efforts of Gorbachev, together with the facts and logical arguments, he could discomfit the most forceful 'shock jock'.

Making his way through various academic thickets, he attended Indiana University, Columbia and Princeton, ending his academic career as Professor Emeritus of Russian and Slavic Studies at New York University. His many books included *Soviet Fates and Lost Alternatives* (2009), *Sovieticus: American Perceptions* and *Soviet Realities* (1986), *Rethinking the Soviet Experience* (1985), and *War with Russia* (2019 – see *Spokesman 142*). However, probably the most influential and provoking work was *Bukharin and the Bolshevik Revolution* (1973). A close aide to Gorbachev stated that the book had a profound influence on Mikhail Sergeyevich, and that there was a possibility of another road for the Soviet Union. We know of course that, owing to the residue of Stalinist-style thinking at the highest levels of the USSR, Gorbachev's hopes for a democratic socialist transition of the Soviet state were thwarted, with disastrous consequences for the ordinary workers and peasants but enrichment of the oligarchs. During the Gorbachev period a friendship between Cohen and Gorbachev blossomed, as did the close relationship between Raisa Maksimovna, Gorbachev's wife, and Katrina vanden Heuvel, Cohen's second wife. Cohen was invited, together with his wife and daughter, to watch the May Day parade from the top of the Lenin Mausoleum.

Cohen must have looked on with dismay the dismantling of the Soviet Union and the emergence of the Commonwealth of Independent States, with the opportunist Yeltsin seizing final control of the other 'White House' with artillery and tank shells. In his book *Failed Crusade: America and the Tragedy of Post-Communism (2001)* Cohen described the terrible consequences for the mass of the population. Whilst the oligarchs accumulated vast wealth and power, the ordinary Russian lost her savings, her pension, sometimes her job and even her life. In Cohen's view the Yeltsin period had to end, some level of Russian financial probity be restored, the ordinary people's earnings needed to be improved, and order restored amongst the competing oligarchs. Economic confidence had to be rebuilt, and Russia had to regain respect on the international stage.

Vilified by much of the media as a Putin sycophant, Cohen stood his ground. Using his vast knowledge of the Soviet Union and Russian history, politics and people he tirelessly explained both the misreading of Russian statecraft and its desire to be on good terms with America and the West. He stated that the arms race between Russia and the US, and the unilateral abrogation of nuclear treaties by the United States, made the world a much more dangerous place. With NATO now on Russia's doorstep and Russia involved in three proxy wars in Syria, Georgia and Ukraine, Cohen feared that the nuclear menace was so close that it was in fact more dangerous than the Cuban Missile Crisis. Not only was there the threat of Russian-American antagonism, but there was the added danger of the joker in the pack, the terrorist jihadist, itching to get his or her hands on some fissile material to spray around Manhattan, Trafalgar Square or the Eiffel Tower. In his 80s, Cohen was not afraid to make himself unpopular: he was brave, an astute teacher, intelligent and knowledgeable. For example, he had been a member of the group advising President Reagan at the Reykjavik talks with President Gorbachev, which nearly ended in the abolition of nuclear arms. Cohen particularly lamented the current lack of serious discussion between experts with contrary views. Under Presidents Clinton and Obama such dialogues were wound down, and under Trump they were non-existent.

The loss of Stephen F. Cohen is a loss for all who want peace and a fairer world. In his final book *War with Russia? From Putin and Ukraine to Trump and Russiagate,* the Afterword concludes:

> Is any leader visible on the American political landscape who will say to his or her elite and party, as Gorbachev did, 'If not now, when? If not us, who?'

*John Daniels*

# Spokesman *Classics*

### The Politics of Community Action
Jan O'Malley
First published in 1977, The Politics of Community Action focuses on community struggles in a small neighbourhood in West London over a period of eight years from 1966 to 1974. So, 50 years on, is it still relevant?
Price: £12.99
184 Pages | A5 Paperback
ISBN 978 0851 2488 75

### Women's Liberation & the New Politics
Sheila Rowbotham
This text was written in the Summer of 1969. A lot has happened since then! The movement which started to grow after the first Women's Liberation conference at Oxford early in 1970 made a significant impact. Fifty years on, 'Women's Liberation and the New Politics' still has much to teach us.
Price: £4
30 Pages | A5 Paperback
ISBN 978 0851 2489 12

### Women's Liberation in Labour History
*A Case Study from Nottingham*
Jo O'Brien
First published as 'Spokesman Pamphlet Number 24' in the early months of 1972, *Women's Liberation in Labour History* discusses the economic, social and political roles of women and children in working class life in Nottingham and elsewhere in the nineteenth century.
Price: £4
24 Pages | A5 Paperback
ISBN 978 0851 2489 43

www.spokesmanbooks.com/acatalog/Spokesman_Classics.html

# The Claims of Women

*Kate Amberley*

*Kate Amberley became President of the Bristol and West of England Women's Suffrage Society in 1871. As outlined here, she created a Ten Point Plan, demanding that the '...same wages should be given for the same work' and 'that there should be no legal subordination in marriage'. Her article was originally published in the* Fortnightly Review *the same year. It was based on a lecture she had given at Stroud in May 1870. Her son, Bertrand Russell, was born in May 1872.*

*If the education is obtained, I do not see why the pay should not follow*

Great authorities have spoken, from time to time, in favour of equality of the sexes; but argument, however logical, falls so powerless when it is met by the ponderous battery of feeling that we must try to enlist this great engine on our side by showing those women who are at present contented, how great is the misery to which the present state of society can give rise. I will endeavour to show them in what way they can assist their less fortunate sisters, and thereby hasten, not, alas, a millennium, but at least a time when every woman will have free scope to cultivate and employ all her faculties and energies, and will be further taught that it is her duty to cultivate them, and a time when, in the eye of the law, she will be the equal of man. Let us bear in mind, when tempted to turn in disgust from the consideration of these claims, for fear that the lovely ideal of woman as she is would disappear were they granted, that 'the useful is noble, and the hurtful base'. As I have said, it is not possible to meet and convince some of our sensitive friends on the field of logic; let us try to meet them on their own ground, namely, that of feeling. I shall appeal more especially to women; for if this battle is to be won, women must be roused from their indifference. When they are united on the subject, the opposition on the part of men would soon cease.

How does an ordinary man of the world answer when he is asked if he is in favour of women voting. He does not say, 'I am afraid of their influence in the elections: they would all be Tories'. He does not say,

◀ Bertrand Russell kept this portrait of his mother

'It would subvert the political and social order of things now existing: they might all be Radicals'. No; he generally smiles benignly and says, 'I do not think ladies wish for it'; and turning, if he can, to some pretty, doll-like girl, he will appeal to her to confirm his statement; which I regret to say she usually does, and he considers the matter settled. 'Why should such fair angels be converted into political drudges?' he will say; and yet till all, or at all events a large number of them are ready to claim a larger share of freedom, we can hardly expect the mass of men to give up the exclusive right to those privileges which they now possess. The history of each reform tends to show us that no class will ever give up any advantage or privileges it may have without a pressure from without. Let the question be political reform, or abolition of slavery, or religious equality, we seldom find those interested in maintaining the abuses clear-sighted enough to help in their removal; and the line of opposition they generally pursue is to descant on the incapacity of the aspirants to power. The distinguished American preacher, Theodore Parker, points this out in his discourse on the Public Function of Woman. He says, 'You know what haughty scorn the writer of the apocryphal book of Ecclesiasticus pours out on every farmer "who glorieth in the goad", every carpenter and blacksmith, every jeweller and potter. "They shall not be sought for", says this aristocrat, "in the public councils; they shall not sit high in the congregation, they shall not sit in the judges' seat, nor understand the sentence of judgment; they cannot declare justice." Aristotle and Cicero thought no better of the merchants: they were only busy in trading. Miserable people, quoth these great men, what have they to do with the affairs of state—merchants, mechanics, farmers? It is only for kings, nobles, and famous rich men, who do no business, but keep slaves. Still, a great many men at this day have just the same esteem for women that those haughty persons of whom I have spoken had for mechanics and for merchants.'

   We have no right to expect any difference in the progress of this reform, which there is not only one class, but a whole sex, interested in opposing. As in those reforms just mentioned, it is not in reality any more advantageous to the possessors of power in this instance to maintain the inequality. Interest and possession, however, so dim the eye of reason, that it cannot see the greater good which looms in the distance to the disinterested vision. Every argument which has been used during the agitation for reform would apply now. The workingmen of this country must have been tired of being told that they were uneducated, unfit for the franchise, that they were virtually represented, or that their interests were safe in the hands of MPs. I am happy to say, though this language was addressed to them scores of times they did not believe it, and persevered quietly and constitutionally till they gained their point. I hope that their access to power will not at once make them conservative in the sense of wishing to keep everything as it has been, and lead them to think that the past, instead of the future, is the thing to rest on and live in. Interested motives, contempt, general dislike to change, and fear of competition, certainly enter into the objection felt by many men for this equality of the sexes, but do not exhaust their reasons against it. There is a large class amongst men who would be justly indignant at any mercenary idea being attributed to them, and who object on the score of sentiment. They dread the disappearance of the gentle, loving, yielding woman, and seem to expect the whole fair sex to be turned into unfeminine monsters.

Mr. Mill might be a magician in the dark ages, to judge by the terror often expressed of the effect of his wand. These alarmists seem to think that, should his incantations succeed, Rip Van Winkle might now take his long sleep, and on waking find all the world peopled by male beings. If this indeed were to be the consequence, I should sympathise heartily with his opponents. But have none of us known women who have, from youth up, been educated with their brothers, nurtured in the feeling of complete equality; others who have taken part publicly in the affairs of the day, and even preached in churches, none the less true women for this training? I refer to some in the Society of Friends, who have consistently carried out this idea of human equality, because they considered it as part of the essence of Christianity. Among these, to allay the fears of our sentimental opponents, we might point to one who is a perfect type of woman as she can be; one who had no false shame or timidity in advocating in public all that liberty and humanity dictated; one whose voice has been raised for near fifty years in the cause of freedom and equality, of all races and sexes, in public meetings, in the privacy of her home, in crowded places of worship; one who, clothed in the neat and simple dress peculiar to the Quakers, joins to their courteous, gentle, and loving demeanour which she possesses in a pre-eminent degree, that calm and peace of a mind at rest with itself, that liveliness and even playfulness of a cheerful disposition, that quick and warm sympathy which is one of the cherished attributes of the gentler sex; a loving wife and honoured mother. I am speaking of Lucretia Mott. Hers is a name that will be long remembered in her own country, though little-known as yet in this.

Another reason why some men oppose any step in this direction is that they are so anxious for the comfort and welfare of the softer sex, that they must protect it from the world and all its hardships and competitions; a worthy wish, no doubt, but one arising from a mistaken principle, and which would be of more use if these protectors were enlightened as to the theory of free trade in labour, and trusted to its effects on the welfare of the protected class. Much more might be said of the objections brought forward by men as reasons for these claims being withheld; but we will pass on to see why, as a matter of feeling, the sympathies of all womankind should be with us. I appeal to their feelings; not that feeling is the safest guide at all times, but because until women have undergone some mental training they will be guided more by emotion than by reason. Too often we are told by them, 'I should be no better off if I had my own property or a vote'; or, 'I like trusting to myself better than to rights'; 'What do women want with colleges? why cannot they be happy and quiet at home?' or, again, 'I should hate to be a doctor or a lawyer, women are not fit for it; they had better look after their husbands and children'. The people who argue in this way fail to perceive that in so doing they are only asserting their own happiness, or their own comfort, and are entirely forgetting the thousands, I might say millions, of women who are alone in the world, who have neither parents, nor home, nor friends, nor fortune of their own, and who are driven to seek these for themselves or to die.

Imagine for a moment the case of a slave woman as she used to be in the Southern States of America, surrounded by the inevitable horrors of that degraded state of society; she may happen to be the petted and pampered darling of a fond master, living in luxury and sure of his indulgence; and when the cry of anguish arising from her fellow-slave strikes on her ear she is only annoyed that any harsh sound should disturb her peace, and impatiently exclaims, 'Why cannot that woman be happy as I am instead of complaining and trying to change her lot?' Perhaps the lot of that other was to work incessantly for a mere morsel of food. Perhaps she was past work, and was about to be sold off away from her hut and her children. Was it wonderful she should raise her voice and wish for some change? Was it not rather wonderful that one woman could so selfishly and indolently enjoy life, because the evil she saw working all around did not touch her individually? We are apt to forget that the priest and the Levite who looked and passed by on the other side are not the examples we would wish to follow; let us look, then, at these sores, and see if we cannot aid in binding them up. To descend to details. Have, for instance, these happy mistresses of comfortable homes ever spared a few moments from their bliss to cast their eyes on a report of the Governesses' Benevolent Institution? It is hardly possible, or we should not hear them urge as an answer to this movement that woman's work is at home. Home is very well where there is one; but what becomes of the work of the fifteen thousand governesses who have no money wherewith to get that home? It is to attain that coveted end, to possess either for themselves or for young brothers and sisters or aged parents, that loved home that so many seek employment in the world.

From the way in which some of 'home' advocates talk, an inhabitant from another sphere might fancy it was a free gift to every human being, to be had by wishing for it; instead of a luxury hardly earned by the labour of its possessor or of its predecessors. The remuneration given to women who enter this career, nearly the only one open to them, is a salary varying from £20 to £100 per annum; out of this they have often to keep relations from absolute destitution. The smallness of the pay comes from the market being so over-stocked, often, indeed, with inefficient workers. But where can they get a good education? What else can they turn their energies to? How are they to get bread? The fact that fifteen thousand women are driven to seek work for themselves is argument enough that by opening more professions, more educational advantages to them, we shall not be guilty, if guilt it is, of alluring them away from their homes to the deadly temptations of the outer world. These benevolent institutions and parallel ones are of use, no doubt, in their day; but those who support them must see that their help, generous and useful as it is to individuals here and there, is but a palliation of an evil, whose root lies deeper and must be cured from the foundation to be effectually eradicated.

This is the number seeking work in one direction; but if we turn to the census of 1861 we shall see that there were in Great Britain, in round numbers, six millions of women over twenty years of age. Half of these were wives, widows, and daughters having no occupation, and so, we presume, well off; one million wives of farmers, shop-keepers, &c., and two millions were engaged in independent industry. I think these figures speak for themselves, and that the cause of two millions is not to be overlooked.

This brings me to the saddest argument that can be addressed to people of feeling and refinement on behalf of the rights of women. Could they be convinced that out of the more than fifty thousand homeless women who lead, in the towns of this country, an existence of moral suffering, of abject helplessness and sin, thousands are certainly driven to it by real want, by the absence of any opening for their industry, their energy, and their capabilities; by the cheerlessness, the hopelessness of their lot; by the absence of education which we have neglected to provide for them; could,

I say, women be convinced that this is so, would they again lightly say 'What is that to me? I should be no better off with this, that, or the other?' No, rather would a life of devotion to that cause seem a small gift to atone for the indifference they had ever felt. This is no place nor time to enter into particulars or to prove the grounds for my convictions; but for the sake of those who will candidly consider the subject I will refer to facts adduced by those who have studied it—by Mademoiselle Daubié, in her recent work on 'The Condition of Women among the Poor in the Nineteenth Century'; also to the books by Acton and Parent-Duchâtelet. They have furnished us with such illustrations of existing evils as must appeal to the compassionate feelings of every fellow-woman; and in the struggles depicted in them to get food and occupation she will see the sign that there is need of amelioration in the industrial position of women, and that we require the justice that can remove causes, as well as the charity that palliates effects.

One of the advantages I hope for in the admission of women to political power is that, their sympathies being strong, they will bring their interest and energy to bear on many injustices of social life, and not so readily acquiesce in the idea that these evils must be borne, and that legislation is powerless to make any impression on them. I think we have experienced in a certain degree the fact that when women see evils they set to work practically to cure them in the limited way open to them. We may be proud of the work done by Miss Rye in emigration; by Miss Carpenter in workhouses, reformatories, and Indian education; by Miss Octavia Hill among the dwellings of the poor in London; and by many others. Would not these ladies be qualified to vote for a member, and to judge of the social and political questions of the day?

I have dwelt mostly on the good to be gained by the women of the industrial classes of society; and as they are at least six times as numerous as those of the upper or idle classes, their cause deserves to be heard first, and what is an injury to them should be removed even at the expense of some of the pleasures or seeming advantages which are supposed to be consequences of our actual state of society. I said that the change should be made, if necessary, for the good of the many; but I do not doubt that there will be no exceptions to those who will reap benefit from this equality of the sexes; for, be the woman rich or poor, married or single, idle or working, it will bring her an increase of happiness by raising her as a moral and intellectual being; and in her improvement, how can man as her companion, and man as her child, fail to taste its fruit? In confirmation of this view it is a pleasure to have the authority of so eminent a man as M.

de Tocqueville, whose advocacy is the more valuable as he does not consider that men and women have by any means the same destiny, and consequently he cannot be suspected of partiality. After giving very high praise to Americans, he says, 'If I am asked to what I attribute the singular prosperity, and the increasing strength of this people, I should say it is to the superiority of their women'. He finds the good effects of democracy in destroying to a great extent this inequality of the sexes as it has destroyed other inequalities; and he thinks it has made woman the equal of man in that country. The Americans, he says, have applied to the two sexes the great principle of political economy which at present regulates industry. They have carefully divided the functions of man and woman, so that the great social work may be better done. I must venture to question the second part of M. de Tocqueville's assertion; for I think that, though the Americans are unfortunately behindhand usually on the great free-trade doctrine, in this case they have been better in their actions than their professions, and instead of 'carefully dividing the functions of men and women', they have opened, or rather not shut, many careers to them which used to be considered the sole province of men. Thus in the United States we find nearly the whole Treasury department worked by female clerks; we find many female doctors, female ministers of the Gospel, and even a female judge. As it is more than thirty years since M. de Tocqueville went to America, perhaps in his time the rapid march of democracy was not as much felt in this department of social life as it is now. They are still fighting there for the political franchise, the denial of which is a badge of inferiority and a real grievance which they still share with their English sisters. It is hardly likely that it will continue; a nation of men who really consider the other half of the nation their equals will not long maintain an inequality when aware of its existence. M. de Tocqueville's strong and emphatic praise of American women will be, I trust, some little reassurance to those who dread that any increase of liberty, knowledge, or power must make that dreaded being known as a blue stocking or a *femme savante*.

   Miss Martineau has also pointed out admirably in her 'Household Education' how absurd is the argument that knowledge unfits women for their work, and asks us if we find men attend less well to their counting-house or their shop for having their minds enriched and their faculties strengthened. She gives her testimony to the worst-managed households being those of the most ignorant women. It seems, indeed, so obvious that the improvement in the social condition of any persons must increase their self-respect, their independence, and that if more is expected of them they will produce more, that the only marvel is how the opposite idea should ever have arisen. Woman, as well as her stronger partner, is a human being first, and has the nature, rights, and duty of one; free scope, equal privileges, and the same standard is all that they require. It is not expected that this will turn the world upside down, or that we shall often see a husband put in the position of Hooker, the divine, who, when receiving a visit in 1585, from two old college friends, had to excuse himself in the midst of the discourse as he was obliged to go and rock the baby's cradle, while a series of similar household disturbances brought the visit to a speedy conclusion. That some women neglect, like Mrs. Hooker, their peculiar sphere, has happened before any talk of emancipation took place, and may, no doubt, happen again; but more education generally makes a more intelligent workman, so we shall not expect to find many Mrs. Hookers who, for the sake of my argument, it is right to say, was a very ignorant and uneducated helpmate.

For this improvement in female education we have much to do. The same means of University training should be open to them, and many of the endowments at any readjustment of their funds should be shared by girls. As a practical instance of the disadvantage they now labour under I will mention Miss Pechey, one of the ladies who have been studying at the University of Edinburgh. They were admitted to the University last November with the distinct statement that they should be subject to all the regulations as to matriculation, attendance on classes, examinations, or otherwise. The lectures for ladies were, however, to be given at a separate hour. Miss Pechey fulfilled all the regulations, passed the same examination, and came out third in a class of 236 in chemistry. There were four Hope scholarships for this class, to be held by the first four students. Miss Pechey most naturally expected to get hers; but, wonderful to relate, it was refused to her because the instruction, by order of the University, had been given at a separate hour. An appeal to the Senatus only confirmed this refusal, though it decided that the women were entitled to the usual certificates. Is this fair play? And, again, is it a thing to be proud of that Miss Garrett, an English lady, should have been obliged to go to Paris, and get from a foreign University a degree for medicine refused to her here.

In the opinion of many, labour is undesirable for women; in the opinion of many others, it is unnecessary. But, if both these opinions were true, and even if we made it our object that labour for money were never forced upon women (which is far from being the case at present), an improvement in their general position would still be needed, in order that they might be better fitted to labour in the fields of art, science, sociology, politics, literature, and society, according to the powers and tastes that all admit them to possess. If in this change woman lose some of those hitherto peculiarly feminine attributes, she will have gained others; what she loses in timidity and sensibility, she will gain in courage and endurance; what she loses in intenseness but narrowness of sympathy, she will gain in breadth. If she lose her fervent religious realisations, she will embrace a calmer but not less noble faith. Her attributes may vary a little; but they will still be feminine attributes, or they will not cling to her. What is beautiful in her nature must be true, and what is true need not fear the inroads of any new opinions or new heresies. If these new attributes are untrue, they may have their day, but will die out; and in the experiment we may hope to have elicited some truth as to what this complex feminine nature is really capable of.

Before leaving this part of the subject, I must refer once more to the unsexing argument. Is idleness the one crowning beauty of woman, that work is an object so much to be dreaded for her? Or is it useful work only that is dreaded, or remunerative work, or possibly work that must be performed outside her home? If this be so, how can we tolerate the 779,000 domestic servants that work in Great Britain, for are they not all women working for their living away from home? Is remunerative work the bugbear of our protectors? I fear there is some jealousy of the competition of women on the part of men; for an eminent medical man lately advised ladies to take to pharmacy when they wished to be doctors, and had the requisite education for the profession. Now, in pharmacy or in hospital nursing there is nothing intrinsically different, as to the fatigue of the life or the delicacy of the work, from doctoring, except that the latter requires a higher education, and consequently commands better pay. If the education is obtained, I do not see why the pay should not follow; and why the woman capable of it should not earn her thousand a year when she can, instead of being content with a hundred in the pharmacy, or possibly forty as a nurse. I will agree with my opponents if they say it is not the work done nor the education given that unsexes a woman, but some kinds of labour, some kinds of misery and want of education that unsex her. I believe that those who are afraid of this she-monster, the unsexed woman, are often thinking of such cases as that of the women working in coal-mines, crawling nearly on their knees, with scanty clothing and begrimed in dirt; or perhaps of the sad specimens of female humanity that haunt the police-courts and bad neighbourhoods of towns. If these specimens are in their minds, I will agree with them that a woman can indeed appear most unwomanly; but we must differ again as to what was the cause of this degradation. It is not from work, not from mixing with men, nor with the world, that this change is effected, but from the same causes that deteriorate the men whom they mix with—ignorance, idleness, poverty, recklessness, vice, and the crimes that follow in their train. These are the causes that unsex women, and make a monster of the being who is capable of shedding such a halo of softness and feeling over the frigid world. But it is not the equality of women with men that is responsible for these degraded forms of womankind; and it is precisely to sweep away these results of our present system that I wish to see women in a higher position in the world. Miss Parkes, after saying that she would like to see many more means of livelihood open to women, expresses a hope that it will be but a temporary arrangement; and that the idea will never be established that women can shift for themselves, and thereby make men less mindful to provide for the women of their family.

Mr. Theodore Parker says that the large class of unmarried women is peculiar to classic and Christian civilisation alone, and that in Christian countries this class is increasing rapidly, and to them the domestic function is very little, often nothing. He does not think that this state will last, as marriage is necessary to the soul and body of man; consequently he hopes this is a state of transition from the time when every woman was a slave and dependent on some man, to a state of independence, where there will be no subordination, but the two will be co-ordinated together. I cannot be so sanguine as to imagine that the balance of the sexes will be so equal, but the temporary stage is one of such great suffering, and must be of such long duration, that we are bound to do all in our power to alleviate it, not dimming our eyes to the facts as they at present exist. We have neither harems nor Mormon homes, on the one hand, for the surplus women who look to matrimony as an occupation; nor, on the other hand, the convents of the Roman Catholic countries, which so admirably filled in their day that longing felt by women for work, a home, a noble life, and devotion to a cause.

It is not work, then, in itself which unfits people for their proper function in life, but a work that is not adapted to their capacity. Perhaps it will be said that this is all very well for the poor, but that ladies in the upper and middle classes of life are the ideals of what women should be, and that you will have no iconoclasts breaking this beautiful image which we have hitherto worshipped. The lady, par excellence, is then to be kept from work and the world, to preserve all those maidenly and matronly charms which are so much prized by men; in plain English, the rich woman is to have no profession but marriage offered to her, that those who do marry may be of the stamp hitherto approved by men. Granted that this is their first profession, what is to become of their energies before this happy crisis, a period often of some duration, owing to the difficulties in the way of early marriage? What is to become of the childless, of the widows, of the spinsters? Are all these to sit at needlework, and dawdle out their day visiting, reading without purpose, and envying their happier companions? But I will not admit that even for the wife and mother a limited sphere of action is desirable, or that the Greek wife described by Mr. Grote in his book on 'Plato', is to be our ideal. Mr. Grote says:— 'We must remember that the wives and daughters of citizens were seldom seen abroad; that the wife was married very young; that she had learnt nothing except spinning and weaving; that the fact of her having seen as little and heard as little as possible was considered as rendering her more acceptable to her husband; ... that her sphere of duty and exertion was confined to the interior of the family. The beauty of the woman yielded satisfaction to the senses, but little beyond.'

Can we wonder, if this was the kind of companion destined to engross the affections of men, that her destiny was a failure, and that among the Greeks marriage was looked upon as essentially commonplace, and that, as Mr. Grote tell us, the wife was quite unable to call 'that pitch of enthusiasm which overpowers all other emotions, absorbs the whole man, and aims either at the joint performance of great exploits or the joint prosecution of intellectual improvement by continued colloquy?' Where the mental inequality exists unfelt and uncomplained of, it is generally because the great aims and intellectual improvement have disappeared under the deadening influences of perpetual contact with a commonplace mind. It is strange that there should not be a greater difference between the lives of women existing at such different periods and in such different surrounding circumstances; for in the life of the married lady held up to us often as the ideal type there is much similarity to the Greek picture we have just considered. She has much spare time. If she live in luxury and wealth, servants, nurses, schools, governesses, in fact, all that money can give, take all the small cares and duties of household life off her hands. Let us recall the old hymn which tells us who finds work for idle hands to do. Her mind is empty, her hands are not required to work; there are no great interests for her; and she is doomed to the life of inactivity, mental and physical, which is thought fitting by the public opinion of her class. Could not this ample leisure be employed in some political, scientific, or social work, according to her aptitudes?

Any attempt made or felt by women to be desirable in the direction of new work is too often crushed in the bud by that fatal advice inculcated so strongly in the education of most girls; that the highest merit of woman is not to be spoken of for good or for evil. A glorious contradiction was given to this theory of womanly excellence in the past, when the first female martyr died for her faith in the sight of thousands of spectators. And to come to modern times, is the life of the Queen one that is devoid of the great interests of political work, official work, and social work? Has this constant public career, these public ceremonies in which she was the central figure, this cultivation of mind which she brings to bear on the duties she has to perform, made her one whit less a real woman, a loving mother, a sorrowing widow, and a ready sympathizer with all forms of sorrow and suffering that come to her notice? She has been held up as a model for English women; and that this has been done shows that the beauty of her domestic life has not been impaired by the public life she has led, and by the great national interests that she has made her own, nor by the shouts and acclamations of multitudes who always rush to welcome her wherever she appears.

I trust now that the time is passed when idleness is a thing desired, and that work will evermore be looked upon as a good thing in itself. If needful and good for the mass, surely each one is the better for contributing to that general good. Remember a saying of an ancient philosopher:—'What is good for the swarm, is good for the bee'. Taking now for granted that the deterioration of women will not be the effect of this change, let us see how materially they can benefit. In the case of the married woman, the right to her own property and earnings will be a great boon in unhappy unions. Where the marriage is happy there will be no need for interference on the part of the law; and, except as regards property, guardianship of children, and divorce, on the same conditions for both, the law cannot enforce equality in marriage; the rest must be left to the private arrangement of each couple, and enters into another sphere.

In the case of those desirous of being married, but waiting many years from absence of sufficient means, surely the woman would be happier, better off, if she were able to employ that time in amassing money to hasten the end in view, than pining in idleness through the best years of her life. The original cause and the cause of the long duration of the present subordinate position of women has been sometimes, and perhaps truly, considered from the Darwinian point of view to be owing to the struggle for existence in which the weaker beings must always be subdued, and, in some cases, even trampled out; and it is asserted that our position is only an effect of that law. Let us grant it is so, for the sake of argument. In former times force was the only power in the world. As civilisation has gone on, the heroic and military virtues have given place to the more amiable ones, and each age has had its type of virtue. The present type is one to which woman can aspire as well as man. In the age of military heroism a woman could seldom distinguish herself equally, though Joan of Arc was an exception to the rule; the field was generally given up to the physically strongest. With the growth of the more refined virtues she took a higher place, and struggled into intellectual existence. That she has been fit for this existence cannot be denied by those who make possession a test of fitness, nor by others who look to the fruits of this moral development of woman in such instances as Mrs. Fry, Madame Guyon, Hannah More, Madame Roland, Madame de Stael, Mrs. Somerville, Miss Carpenter, Miss Cobbe, and an innumerable host, who have done work in various spheres.

Now that Government, the arts and sciences, have all thrown a gentler glow over the aspect of modern life, the sphere of action is still further enlarged, and by her power to gain admission to that which is still withheld from her, she will prove her right to political existence even to the mind of a thoroughgoing Darwinian. That this emancipation may be accomplished, it is certainly necessary there should be some struggle, in the sense of agitation and of efforts made to remove existing grievances; hence have arisen societies, discussions, bills, debates, and petitions. The more women help in this agitation the sooner will they all start fair in the race; and till women have votes, those things which concern them alone are sure to be made to give way to those that are pressed for by the constituents who have power at elections. If we believe in representative institutions we must be convinced of the material advantage to women in the acquisition of a vote. Political power is a protection, and it is in that sense, and not as a right, that we demand it for women. Much has been done in the last few years in the education of the public on this subject through the means of literature, the press, and speeches. We often hear a feeling expressed that a polling-booth would be no fitting place for a lady. But I confess that my experience teaches me that a polling-booth, in the early morning, is a far quieter place than the hustings on nomination day, or the public market-place on polling day; yet these two places have been sanctioned by public opinion as quite suitable for the wives and friends of candidates at elections. Even if it be a little unpleasant, a trifle noisy, and if an occasional flour-bag or egg finds its way to the sacred precincts, the lady is supposed to be able to bear it with equanimity for her husband's sake. I ask her to go through much less inconvenience for her own sake and that of her country.

Mr. Mill first proposed to give the political franchise to women in 1867, when seventy-five members voted for it in a house of two hundred and seventy-three. Amongst the seventy-five were found two most distinguished members of the present Government, Mr. John Bright and Mr. Stansfeld. Mr. Jacob Bright brought in his Bill for removing the electoral disabilities of women this year, and the second reading was carried by a majority of thirty. On the motion for going into committee it was, however, thrown out. The ignorance of women of any movement in the world of politics is often mistaken for apathy; but once show them that any practical grievance would be attended to at once, if they were the holders of political power, and had authority to question their members and make their vote depend on his answer, and they would become as keen politicians as the men they live with. No sooner was the municipal vote granted to women than, in the small town of Leicester, between two and three thousand put themselves on the municipal register. Does this look like indifference? An immense step has just been taken by women in connection with the election of School Boards. The keen interest taken by them in this matter of education, led not only to their voting, combining, and canvassing, but also to their becoming candidates. Miss Davies and Miss Garrett, M.D., owed their wonderful success in a great degree to the desire felt by women to entrust their interests to one of their own sex; but as women alone could not have made up the 47,000 votes given in Marylebone for Miss Garrett, we have a sign that men have no objection to be represented by a woman. The elections were carried on so much on the plan of the parliamentary elections, that it has served as a test of the capacity in women for the function of voter or candidate, and must have disposed of the objections urged on the score of incapacity and unfitness. It is to be hoped that women all over the country will come forward as candidates for School Boards; and by the way in which they carry on their contests, will show that in assuming this public attitude they do not adopt the well-known male electoral vices, but bring the feminine virtues they are credited with to their aid. If this be so, and we have less lying and humbug, less treating, less intimidation, less unscrupulousness, and less rowdyism, with our female candidates, we shall all rejoice in the day when women began to share the wider interests and larger life of men.

Great steps are also being taken in the higher education of women; a college, giving the same education as that of Cambridge, has been opened at Hitchin, and various courses of lectures in London and Edinburgh by University professors. This, however, is too large a subject to enter on here. Many, moreover, are willing to go with us in the education question, but say that, when education is what they call finished, women should not seek to go further through the golden gate of knowledge. But to expect a woman to cultivate all her faculties and her mental powers, and then to acquiesce in a life of inactivity, 'a life of nothings, nothing worth', would be unreasonable. If she asks that the same opportunities be open to her as to men, her demand is only just and rational; nor is it a loss to herself only, but the community also sustains a loss of force, labour, and energy by barring the door to every external occupation to one half of the human race. Have these timid people so little faith in nature, so little faith in their power to win a woman, or in her instinct to be a true wife and mother, that they must hedge this rebellious creature round so that she may have no outlet except into matrimony? When she enters into it, cannot you trust her to find out how far it engrosses her whole being. Precisely in proportion to their enlightenment will women on the whole see more clearly what their true work is, and that that work need not always be identical with that of men, nor yet so opposed that the men must work, and the women must weep.

With progress, men break into varieties in their employments. Why should not women follow the same law of social economy? When all is open to them, it does not follow that they will become soldiers and sailors, iron-workers and blacksmiths. The law of natural selection will operate here as throughout creation, and what they are fitted for they will perform; if they do not perform it, they will soon be replaced by the fitter instrument for that particular work: but in the new order of things we have never yet had fair trial. It still remains to be seen if they cannot fulfil the offices of doctors, of preachers, of educators, of clerks, of poor-law administrators, of printers, of reporters, of shop-keepers, of book-keepers, as well as these offices have been filled already. The relief of the poor has been considered as a fit sphere for woman for some time, and the reason that, amidst all the good they have done, there has been so much mischief in charity, is the absence of the sense of responsibility in their dealings, and their ignorance of political economy. These two defects would, I hope, be remedied in a state of society such as we wish for. It will be noticed that in the spheres of work I have indicated for women, I have not mentioned any that can be objected to on the physiological ground that long-continued muscular exertion is injurious to them.

Having now passed in review, certainly in the most cursory manner, the various obstacles to those changes, both in public opinion and in the law, which we, who are supporters of the claims of women, are anxious to effect, I will briefly sum up the measures which we conceive to be required in order to secure that equal justice, which is all that we demand.

- 1st. We desire that there should be a great improvement in the education of girls, and a restoration to them of those endowments originally intended for both sexes, but which in some instances have been appropriated exclusively to boys.

- 2ndly, and as a natural sequence to the first requirement. That equal facilities should be granted to women for the attainment of the highest education and of University degrees, in order that their special faculties may not be consigned to compulsory idleness, but may be turned to the benefit of society.

- 3rdly. That all professions should be open to them, and especially that no new Act, medical or otherwise, should actually exclude them as they are excluded now.

- 4thly. That married women should no longer be debarred from the separate ownership of property, on the same terms as married men.

- 5thly. That a widow should be recognised by law as the only natural guardian of her children.

- 6thly. That the franchise should be extended to women as a means of power and protection in all matters affected by legislative action.

- 7thly. That political and social interest and work should be open equally to them, so that if there be talent or aptitude in any of them the State may not be the loser, alike by the exclusion of those qualities which they share with men, and of those which are characteristic of themselves.

- 8thly. That public opinion should sanction every occupation for women which in itself is good and suited to their strength.

- 9thly. That there should be no legal subordination in marriage.

- 10thly. That the same wages should be given for the same work.

But I hear some of you ask—'All this being granted, *cui bono* and I answer you simply—We hope and wish to try if an infusion of justice, of new vigour and new life, of warmer sympathies and larger hopes into women's lives, will not alleviate some of the suffering of this struggling life. If it alleviate but few, it is well; if it have the effect I anticipate, it will do more. At all events let us hope. *'Die Welt wird alt und wieder jung, Doch der Menschhofft immer Werbesserung!'*\* and the day will be sad when we become sceptical of individual and social progress.

In conclusion allow me to refer to the chapter on the family in a beautiful book on 'The Duties of Man', by the great Italian patriot, M. Mazzini; where, after dwelling on the beauty of family life, he says:

'Like every other element of human life, it is of course susceptible of progress. Seek in woman not merely a comfort, but a force, an inspiration, the redoubling of your intellectual and moral faculties. Cancel from your minds every idea of superiority over woman—you have none whatsoever. Long prejudice, an inferior education, and a perennial legal inequality and injustice have created that apparent intellectual inferiority which has been converted into an argument of continued oppression. Man and woman are varieties springing from the common basis—Humanity. There is no inequality between them, but— even as is also the case among men—diversity of tendency and of special vocation. Are two notes of the same musical chord unequal or of different nature? Man and woman are the two notes without which the human chord is impossible. Consider woman, therefore, as the partner and companion, not merely of your joys and sorrows, but of your thoughts, your aspirations, your studies, and your endeavours after social amelioration. Consider her your equal in your civil and political life. Be ye the two human wings that lift the soul towards the Ideal we are destined to attain.'


\*The world is getting old and young again, but people always hope for improvement!


*The portrait of Lady Kate Amberley was drawn by her brother-in-law, George Howard, Lord Carlisle, in 1864.*

# Reviews

## Women's Co-operative Guild

Ruth Cohen, *Margaret Llewelyn Davies: With Women for a New World*, Merlin Press, 2020, 256 pages, paperback ISBN 978058367591, £17.99

The Women's Co-operative Guild which, sadly, came to an end in 2016, played a key role from the 1880s onwards in awakening working class women to the task of fighting for their just rights and for gender equality. Although many outstanding women were involved in its achievements, none did more to inspire the struggle than Margaret Llewelyn Davies, its General Secretary from 1889 to 1921.

This book is a biography of Margaret which is not only a detailed record of her administrative and public work but also an account of her family background and personal life, touching on the lives of many of her distinguished relatives, friends and acquaintances.

Perhaps unexpectedly, Margaret came from an upper middle class family. She was the only daughter among the seven children of the Rev. John Llewelyn Davies, the incumbent of Christchurch Marylebone, in West London, a theological scholar who had translated Plato's *Republic*. Her brothers were educated at public schools and universities and she herself was well educated, finally at Girton College Cambridge, of which her aunt, Emily Davies, was one of the founders.

Margaret's father was a Christian Socialist influenced by the theologian, F. D. Maurice, and her mother, Mary Davies, was well educated and came from a Unitarian family. Other relatives were linked to progressive causes and were highly knowledgeable.

However, Margaret's mother insisted that she should return home before completing her course at Girton, probably for health reasons, and she took over some of her mother's responsibilities. Influenced by the family tradition of public service, Margaret helped to run two youth clubs.

In 1886 she read *On Labour* by William Thornton, which led her to join the local co-operative society and, in December 1886, she attended a meeting of the local branch of the Women's League for the Spread of Co-operation. This organisation had been founded in 1883 by Alice Acland, a vicar's daughter, and Mary Lawrenson, a teacher. Margaret was deeply impressed and accepted the position of branch secretary. So successful was she in recruiting members and arranging attractive topics for discussion that she was soon elected to the Central Committee of what was now called the Women's Co-operative Guild. When, in 1889, Mary Lawrenson tendered her resignation as General Secretary, Margaret was put forward and was elected to take her place.

This all happened about the time her father was pushed to resign in Marylebone and accepted a living at Kirby Lonsdale in the Lake District. Margaret moved with her parents and took the opportunity to visit most northern industrial towns to build up the Guild branches there. She did this by arranging topical talks and lectures and insisting on educational objectives. She argued that Guild meetings should not become 'mere mother's meetings'. The wives of skilled workers, who kept the home while their husbands worked, were attracted to join and the membership grew rapidly. However, Margaret was very conscious of the plight of the poorest women and worked hard to draw in and assist women in the most depressed classes.

One of her basic objectives was to provide women with the knowledge and the will to speak at meetings and seek election to Co-operative Society boards and committees, which were totally dominated by men. This did not enhance her standing with male board members, but she was undaunted. She fought relentlessly for women's rights and won Guild members over to support her.

She battled for cheaper divorces and argued for divorce by mutual consent, which was opposed by many members on religious grounds. She campaigned for improvements in Lloyd-George's National Insurance Bill and helped to achieve mothers' allowances. She sought to achieve better wages for low-paid women, supported trade unions, and organised assistance for workers on strike. She devoted immense amounts of effort to campaigns to give women the right to vote and argued for electoral reform to give women and men votes on equal terms.

Margaret was an internationalist and helped to found the International Co-operative Alliance. She was a pacifist and opposed the First World War, which divided the Guild membership. This led, eventually, to the defeat of a resolution at the Guild's national conference calling for peace negotiations and no annexation.

Despite this, Margaret stuck to her beliefs. The result of her work over the years was reflected in the growth of membership. In 1921, the Guild had nearly 50,000 members organised in nearly 800 branches. To the consternation of the membership, Margaret decided to resign in 1921. Apart from the reaction within the Guild, the United Co-operative Board, with whose members she had previously clashed, invited her to chair their Co-operative Congress at Brighton in 1922 – the first woman ever to do so. Subsequently, she was also granted the Freedom of the Guild with a testimonial of £700.

Margaret lived on to 28 May 1944, along with her former assistant secretary, Lilian Harris, with whom she finally moved to Dorking in Surrey. She suffered from bouts of ill health but, in July 1933, spoke at a huge Guild Congress to mark the 50th anniversary of the Guild's foundation. She declined to write the history of the Guild but edited two books recording the lives of working class women: *Life as We Have Known It* and *Maternity*. She continued to take an interest in many progressive causes, to which she made bequests in her Will, and remained a deeply committed pacifist to the end of her life.

Ruth Cohen's biography is a magnificent record of the life of a truly outstanding woman who was an inspiration to the Left in Britain and particularly working class women. Not since Jean Gaffin and David Thomas published *The Centenary History of the Co-operative Women's Guild* in 1983 has there been a publication which conveys the tremendous achievements of the Guild as this book does. It is a fascinating, enlightening and inspiring volume. Feminists, co-operators and activists on the Left, apart from readers generally, will benefit greatly from reading it.

Stan Newens

## Liverpool at home

**Matthew Thompson, *Reconstructing Public Housing: Liverpool's hidden history of collective alternatives*, Liverpool University Press, 2020, 408 pages, paperback ISBN 9781789621082, £25.95**

Matthew Thompson's new book is a comprehensive and very interesting contribution to a Leverhulme Trust initiative which funded his three-year research project: *'Reimagining the City: New Municipalism and the Future of Economic Democracy'*. Amongst Leverhulme's aims are:

> A focus on new municipal initiatives to incubate and stimulate a new co-operative economy – specifically the development of institutional ecosystems for worker owned co-ops and other forms of democratic enterprise, in cities contending with the adverse conditions of global economic restructuring and neoliberal austerity.

I was especially interested because I have also recently been looking through papers relating to the history of economic development in

Sheffield and South Yorkshire during the 1980s, in which I was involved through the City Council and through the Centre for Local Economic Strategies (CLES). I had previously lived in Liverpool, during the late 1960s and early 1970s.

Matthew Thompson reminds us that Engels had little praise for either city as he travelled around the country, condemning the appalling 'cellar dwellings' in Liverpool and smoke filled air in Sheffield. In common with the overall aims of Leverhulme Trust, Thompson describes the aim of his book as 'trying to see in collective alternatives a "real utopian" potential to transform public housing and urban economics'.

Thompson re-uses the word 'commoning', recalling the role in England of common areas used by communities for grazing animals, and the village green for festivals and common use. The 'Diggers Song', originally told by Gerrard Winstanley and arranged by Leon Rosselson, reflects the utopian view:

'In 1649, on St Georges hill, a ragged band they called the diggers came to do the people's will. "We come in peace", they said, 'to dig and sow; we come to work the land in common and to make the wasteland grow. We are free men though we are poor. This world was made a treasury for everyone to share. You Diggers all stand up for glory stand up now!'

I was reminded of local research here in Sheffield where a plot several acres in size was simply known as 'Common Piece' before being acquired by a local steel-maker following the 'Enclosure' legislation, and then used as 'quality' housing for himself and other key operatives.

Thompson's research exposes the totally unsatisfactory state of housing in Liverpool during the 1960s/70s: multi-occupancy and poverty of the Granby street area, between Toxteth and the University, matched by the other unsatisfactory alternative of re-locating swathes of residents and tenants suffering appalling poverty in housing miles away from Liverpool in Skelmersdale, Halewood or Kirkby. It's useful to be reminded of the significant and positive role played by Shelter and Des McConaghy in setting up the Shelter Neighbourhood Action Plan (SNAP) – an action research project which actually visited and talked with families about their financial needs, and possible futures, rather than being simply on the receiving end of housing new build or restoration decisions by a paternalist local authority. This important movement, along with the Home Office community development projects, offered a way forward through collective housing co-operatives, for which the city became well known.

A number of varied examples are described in detail: Granby Street, Weller Street, co-operatives on the Kirkby estate in Knowsley, and around Eldon Street, established by co-operators he describes as 'Eldonians'. These efforts were taken up by Harold Wilson's Government through Housing Mnister, Reg Freeson. Freeson's adviser, a Liverpudlian called Harold Campbell, set up a working party on co-ops, which recommended they register with the Housing Corporation as Housing Associations so that, through the 'fair rents' regime, housing became affordable for those on low incomes for the first time. The book illustrates how Liverpool's experience of co-operative housing linked with the movement away from council tenure as housing associations took hold.

Perhaps Matthew Thompson could have explained more of the historical tradition which had led to decades of Conservative rather than Labour control of Liverpool City Council, despite the example set by Bessie Braddock, MP for Liverpool Exchange from 1945 to 1970, and her husband, Jack, who was also a Labour Councillor.

Surprisingly, Liverpool went Liberal in 1972, which offered the opportunity for this wide variety of co-operative pilots to become established. Thompson's detailed analysis exposes the problems which the Liberal pursuit of 'choice' brought into tenants' movements at the time. He briefly mentions Kirkby, where women made an input into practical solutions to their housing issues, and suggests rightly that housing policy was a key factor in allowing Militant — a very male dominated group which used its opposition to housing co-operatives, in favour of municipalisation in the name of socialism, to gain control of the District Labour Party, and thence Liverpool City Council in the later 1970s. I recognise the author's description of a political culture of 'anarchic syndicalism' from when we lived there, during the period from 1968 to 1974. This contrasted sharply with Sheffield and its solid Labour and trade union solidarity, which dominated throughout its virtually continuous control of the local authority since 1926.

This takes the conclusions of the book, politically, down the road of current assumptions about land and property. Individual owner-occupation, rather than collective or co-operative housing models or indeed private or public sector tenure, are jealously guarded by conservative economists. They celebrate this because of added status and prospective wealth it has offered, during decades of steep house price inflation, to owner-occupied living and associated consumption, with people living in one, two or three car families. Thompson's conclusions are well worth considering, particularly in light of continuing social inequality, climate changes, and health pandemics.

I was left wishing for a more gendered and racially based interpretation of Liverpool's housing story, especially that the role of women in tenants' movements was more fleshed out. Perhaps a further study might offer some insights along these lines.

*Helen Jackson*

## Walking the tightrope

Ali Smith, *Summer*, Penguin 2020, 386 pages, ISBN 9780241207062, £16.99

'You carry your own joy with you wherever you go', wrote Kurt Schwitters in a letter to his wife Helma in April 1941 from Hutchinson Camp at Douglas on the Isle of Man. Schwitters' story, and those of the painter and memoirist Fred Uhlman and their fellow German and Austrian internees – some 1,200 in total – occupy a central place in the last of the novels that make up Ali Smith's mesmerising *Seasons Quartet*, captured, as it were, in a blaze of summer light. Schwitters' release was agonisingly delayed, but he enjoyed the immense good fortune of having washed up among a crowd of like-minded intellectuals and creative spirits in an environment, glowingly detailed by Smith, that was uniquely well-disposed among wartime camp regimes to the practice, however circumscribed, of his particular art: *Merz*, or 'the combination of all conceivable materials for artistic purposes' ... 'an existential project that seeks to lend shape to hapless circumstance' (Roger Cardinal).

Of course, 'A prison is a prison' even for Daniel Gluck, the centenarian magus or Pied Piper of 'arty art' and its lifelong powers of liberation first seen in *Autumn* but here recounting his own journey, and his father's, to Hutchinson as a grim litany of dispossession and tabloid-inspired hostility. Once disembarked, however, as on Prospero's island, the faery wand of Schwitters' porridge sculptures and collages assembled from canteen waste, Uhlman's ink drawings of a child's balloon floating high above the fences and war zones' 'heaped-up hills of skulls' (Mandelshtam on another historical apocalypse comes inexorably to mind) and figures, sacred and profane, from myth and fable cut into the paint on blackout windows to let the light in – works its magical transformation. It's as if an entire world comes together, intoxicated like Smith herself (always!) by the imaginative impulse, in a sharing of artistic projects, gifts, performances and Dadaist happenings light years from the 'coerced community' or

*Zwangsgemeinschaft* of Adler's classic work on Theresienstadt which Daniel appears to refer to, with characteristic playfulness and some prescience, in a letter to his sister in Occupied France. Above all it's the fresh engendering of dialogue, for Smith the *ne plus ultra* of social reconciliation, around simple tokens of esteem – 'There is always a point where a hand reaches out to another hand', she has said – that marks out the enchanted terrain: a Butlin's lapel badge, the one redeeming artefact from an earlier camp, in exchange for lessons in rudimentary English; quality drawing paper from an artist's precious hoard in return for a Heine translation; a bunch of meadowsweet, remembered as a local girl's gift to an internee in a previous war. 'Do you like to see things as they are *and* as they aren't?' Uhlman asks Daniel, in perhaps the ultimate formulation of where the *Quartet* as a whole wants to take us, starting him on his further journey to the guru-like, passionate aestheticism that will make him what one commentator claims is the novels' 'moral centre'. Hutchinson, meanwhile, represents a rare moment of collective enlightenment (encouraged on the ground by a benign officialdom) that not only draws in the 'fine people of Douglas crowded in their hundreds up against the wire', all antagonisms seemingly suspended, to share in the artistic extravaganza, but can also count for its eventual dissolution upon the rectitude of a fair-dealing parliament ('They know you don't put innocent people in prison. The British are just. They're practical. They're forbearing') witheringly counterposed by the shifts in consciousness of the Daniel of 2020, in and out of memory, and grown more cognisant of a systemic malfunction, to the present day: 'Thugs and showmen in power, he says. Nothing new. A clever virus. That's news. The stocks and shares will shake. There'll be people who do very well out of that. One more time we'll find out what's worth more, people or money'.

The stress throughout the four novels, as is well known, has really been on just such a precise, often unforgiving contemporaneity. *Autumn's* opening sentence misquoting Dickens, 'It was the worst of times, it was the worst of times' – the revolutionary ardour, alas, thoroughly discredited – left no doubt about the seriousness, or urgency, of the undertaking, although any reader forewarned in 2016 of the severity of the plunge into crisis upon crisis that followed might very well have responded, like Edgar in *Lear*, 'And worse I may be yet: the worst is not so long as we can say, "This is the worst".' Daniel Gluck, the *Quartet's* first refugee, comes ashore like Odysseus in a swoon of fabulation, dream images, literary and philosophical allusion that takes in artworks by Boubat and Dubuffet, objects of reverence and productive (the former) of an unfolding, complex symbolism, before the intractable, murderous reality of social division and exclusion intervenes to stop him in his tracks: corpses of migrants, asylum seekers, a drowned child whose image went viral, thrown up on the beachhead alongside holidaymakers 'under parasols'. Beyond this lie, in short order, the murder of Jo Cox, the virulent antagonisms and demagoguery of the Brexit referendum and its fallout, and the intensifying focus on the degradations of detention – the one issue that comes, if any does, to dominate Smith's moral agenda from *Winter* and *Spring* onwards. Daniel, stymied, 'looks from the death to the life, then back to the death again', at the unbridgeable gap between victimhood and the complacency that endorses it and can only muster, *sotto voce* as it were, the single, despondent phrase 'The sadness of the world'. How *are* art and literature, for all their mercurial refinements, to respond to the scenes of senseless agony that constantly confront them? In *Artful* (2012), Smith speaks of a 'magical shifting of the position of observed and observer', the way what's witnessed imaginatively rebounds upon the viewer with an implicit call to action – 'you must change your life' – and cites a poem by Szymborska, 'Greek Statue', which cedes primacy to the human content of the artwork, 'dismembered' by time but 'bloody lucky, never having been alive in the first place: *When someone living dies that way / blood flows at every blow'.*

*Summer*, then, might serve as a test case for where Smith's evolving priorities, and ours, currently lie. She wrote in 2015 of forty years' indebtedness to John Berger, his unstinting attentiveness to 'the otherwise drowned-out and inaudible – the voices of the invisible', and published, more or less at the same time, her 'Detainee's Tale', the most harrowing account she's given (Brit the Custody Officer's ferocious testimony in *Spring* notwithstanding) of what it means to be an immigrant on the receiving end of Her Majesty's Pleasure. Here, indeed, witnessed at first hand, are the invisible, Ghanaian and Vietnamese, with their all-but-invisible indictment: 'I thought you would help me' – suddenly the blandishments of the island confraternity seem a long way off. 'I am an idiot. But I'm learning': a pre-2015 Smith and a post-? – anyway, the *Quartet* was to follow. *Summer's* immediate focus, like its seasonal predecessors, is on her own shock troops, a cast of characters yielding to or defiant of the *Zeitgeist* – 'All manner of virulent things are happening', not least a rampant defeatism masquerading as indifference – and on the equivalent of what Robert Schumann called the musician's task 'to bring light into the depths of the human soul'. To the latter end, she draws again upon another invisibility, the work of forgotten or under-represented women artists, 'bold and independent experimentalist[s]' like (Marina Warner) herself, all in a sense collagists, chameleons, modernist forerunners: Boty, Mansfield, Hepworth, Tacita Dean, the irrepressibly versatile Aliki Vougioukaki from *Artful*, and now Lorenza Mazzetti, whose London Diary of the Sixties London film scene — comprehensively paraphrased – is an object lesson in joyous creativity against all the odds a traumatised war refugee can encounter (like the Croatian immigrant Lux in *Winter*, she is 'A girl like a broken bird'). Mazzetti's image of a walker high on a roof's edge as if on a tightrope, and freighted perilously with suitcases, speaks to the hope animating the whole cycle, 'like the blade of a knife', as Smith has said, 'and you are balanced on it … we need to be light as we traverse incredible dark'.

But such a dark! – which even Shakespeare's late romances, here *The Winter's Tale*, with their utopian promise of a refracted society returned to health leave as intractable as ever at the novel's close. *Summer* declines, one feels, to look too far into its depths, at what motivates the drugged and downtrodden signing up to their own expropriation: the evangelist Mercy Bucks' pro-Trump white revivalists, her 'tending to them tenderly and pickpocketing them while she does', or the Wetherspoons drunkard glimpsed 'crawl[ing] on all fours to the seafront with his trousers and underpants down around his ankles'. The travails of homelessness and destitution, apart from those visited on outsiders ('How great it is to eat when you're hungry' is a refrain running through Mazzetti's diary), are left largely unexamined – Berger's *King: A Street Story* is a more compelling guide to *that* invisibility. The propertied, the educated and media-savvy, the artistic, whose tale this chiefly is, nevertheless in nearly every case learn to step down from their wire and engage in the various activisms, political and counter-cultural, that go towards defining what Smith might mean by a true heroism: 'I am not a hero! I am not a masterpiece! But I am a brother' proclaims the detainee rescued from incarceration and oblivion by three of *Summer's* interventionist women, although 'Hero', his name, *is* the last word in the book. Hannah Gluck spirits Jews across the Swiss border from France, Iris the Greenham warrior from *Winter* throws open the doors of her sister's mansion to migrants cynically evicted, because of Covid, from lock-up, delivering in the process the *Quartet's* most explicit, fulminating denunciation of the Johnson government: 'I will shout at the walls and the frontiers to break open. I will keep my nose open for the power-shit', Smith's Berger tribute promised, and she does.

*Stephen Winfield*

## Life in Jarrow

Ellen Wilkinson, *The Town That Was Murdered – The Life Story of Jarrow*, Merlin Press, ISBN 9780850367492, £14.99

Ellen Wilkinson was born in Manchester in 1891 and became the first woman to sit on the City Council. She was a socialist, feminist, and supported women's suffrage. She was elected as the Labour MP for Jarrow in 1935 and she became the first female minister of education in the 1945 Attlee Government.

This book documents the history of Jarrow and the impact of capitalism on our communities. Many people will recognise the name because of the Jarrow March in 1936 when workers marched to London to raise awareness about their plight. It is particularly poignant for me as, in 2014, I became active in the People's March for the National Health Service, which retraced the steps of those brave workers in 1936. They were inspired by the Jarrow marchers – 200 men who stood up to the extreme poverty of the Depression era in Britain. The NHS protest was led by a group of 24 Darlington mothers and I was very proud to play a small part in the work they did. The NHS continues to face the threats of cuts and privatisation.

Ellen Wilkinson traces the rise of capitalism in the North-East region. It opens with a chapter on the outbreak of cholera in Jarrow, quoting from *The Newcastle Courant* in 1832:

> 'One of the most remarkable features it still exhibits. The narrow and dirty lanes in the lower parts of the town, and the confined and ill-ventilated passages which are numerous in the upper, and in which the dwellings of the poor and wretched are situated, have been, with few exceptions, the only places to which the disease has penetrated, and in which it has revelled with all its fatality ... it might also be inferred that it is a malady as far as regards predisposition PECULIAR to the poorer portion of the population.'

Even cholera was somehow the fault of the poor people. Many of us will recognise that this demonisation of poor people continues to this day. We regularly see this blame culture, not only through mainstream media but also across social media.

Early on, Ellen Wilkinson talks about the Jarrow pits and their role in providing work for the community. It is interesting to note that in the early chapters, when we hear about the treatment of the pit workers and the bonds they 'signed', this reflects so closely on the way the new 'gig' in the UK and globally has impacted on workers' rights. It is also quite shocking the level of victimisation against any worker who was identified with the union and the impact that had on their ability not only to get work but also to organise effectively.

Following the defeat of the 1832 miners' strike, conditions for workers were seriously affected and any attempt to maintain wages was undermined. Jarrow became known as the 'slaughter pit'. The demise of the shipping industry followed some eighty years later. Ellen Wilkinson sets out the stark reality of workers during this period. Not only the hard and enduring work but also the inhumane conditions in which these workers were forced to live. As she points out 'Jarrow, of course, was not the only town where insanitary conditions prevailed during this pirate period of nineteenth century capitalism…. where the health of the workers was not even considered where profit was to be made.'

This book is the story of Jarrow, but it is also the story of so many other towns of the time. It tells of the struggle of workers, of trade unions, and of those who had the courage to speak out. It is also the story of today. Ellen Wilkinson stood alongside the workers of the time. We need more people standing alongside workers now.

The past decade has seen a significant decline in the trade union movement. We have seen the casualisation of the workforce and massive attacks on workers' rights and their working conditions. We have seen the number of children living in poverty increase rapidly in the past few years, and we know that many of these young people are in working families. There is a huge increase in families relying on foodbanks to survive. This is the shocking reality of the world we live in.

The pandemic we are currently living through has shone a bright light on the inequalities in our society. The union movement has seen a resurgence in membership and in workplace activity. We can learn much from our predecessors, from their fights for justice and equality. We need to use this history and the current climate to rebuild the workers' movement and to fight for decent pay, conditions and contracts for workers in Britain and globally. We have nothing to lose but our chains.

*Louise Regan*

## On file

**James Smith, *British Writers and MI5 Surveillance, 1930-1960*, Cambridge University Press, 2013, 226 pages, hardback ISBN 9781107030824, £75**

Bertrand Russell's Metropolitan Police Special Branch file number was 405/43/1061. We now know this because of documents published as part of the Undercover Policing Inquiry in London, which commenced its proceedings online in November 2020. The Inquiry spans the years from 1968 to the new millennium, with its first phase addressing the activities of the Metropolitan Police Special Demonstration Squad (SDS) between July 1968 and the end of 1972 'approximately'. In the 1960s, the SDS commenced its surveillance of the Vietnam Solidarity Campaign, and the Inquiry has enrolled Ernie Tate and Tariq Ali, active in the VSC and therefore named in these Special Branch reports, as 'core participants'. With notable thoroughness, some Special Branch reports list associated registry files, and one such list includes Bertrand Russell. It is curious that his file appears to have been opened in 1943, when Russell was living in the United States.

Why read this book? First of all, we discover that it is unusual for Special Branch files to see the light of day. Citing sources at the National Archives, James Smith says '...Special Branch has released only a handful of its records and destroyed much of the rest'. Professor Smith has combed some of the more literary of the five thousand or so 'pieces' that MI5, the UK domestic security agency, has chosen to release from the founding of the Secret Service Bureau in 1909 through to a cut-off point in the 1960s. He explains the structure of an MI5 file, and how the service interacts with the Metropolitan Police Special Branch and the external Secret Intelligence Service (MI6), which does not release any 'pieces'. In so doing, he recounts revealing stories of how some of the spied upon, such as George Orwell and Stephen Spender, came to assist or work with the surveillance agencies.

In December 1945, writing in *Tribune*, Orwell asked, 'when a Labour Government takes over, I wonder what happens to Scotland Yard Special Branch? To Military Intelligence? ... We are not told, but such symptoms as there are do not suggest that any very extensive reshuffling is going on.' The London Undercover Policing Inquiry rather bears out that insight. Smith's book is a highly readable and informative attempt to address

Orwell's justified fears and 'understand how these evolving security-intelligence forces monitored the left-wing writers and artists of his generation'. For it is predominantly the Left that authority perceives as 'subversive'. In addition to Orwell and Spender, those who came under the spotlight include Ewan MacColl, Joan Littlewood and Theatre Workshop; the wider Auden Circle; and Arthur Koestler. Professor Smith has performed a notable service in disentangling their stories as told by Britain's spies.

*Anthony Lane*

## Russell and anti-war politics

**Aled Eirug, *The Opposition to the Great War in Wales*, 1914-1918, University of Wales Press, 2018, 250 pages, paperback ISBN 9781786833143, £24.99**

In the middle of World War I, Russell was politically inspired by young socialists and pacifists from the South Wales valleys who flocked to hear him speak against the war. In later years, for private rather than public reasons, he was captivated by the rugged natural beauty of the country's North, where he died in February 1970 —at Plas Penrhyn, the cottage in Penrhyndeudraeth, Merionethshire, where he lived for the last fifteen years of his long life. He was born 97 years previously at Ravenscroft, his parents' "very lonely" country house on the River Wye in the border county of Monmouthshire (*Auto*. 1: 10). It is sometimes easy to forget how large Wales looms in Russell's biography, far more so than it is to overlook his lifelong commitment to international peace. Anti-war politics in a Welsh historical setting is the subject of Aled Eirug's fine study. Observers of Russell's cardinal political preoccupation would likely agree with this author that an understanding of World War I "is incomplete without an appreciation of the diversity of responses to it, including the opposition to the war" (p. xv). This last dimension as it affected wartime Wales is probed in depth by Eirug, whose project gestated for decades as he pursued a career in journalism that included a long stint as head of News and Current Affairs for BBC Wales. But it has been well worth the wait. Eirug has produced not only a valuable addition to the monograph series of which his book is a part ("Studies in Welsh History"), but also to the historiography of the British Home Front as a whole.

In four lengthy chapters, Eirug addresses, first, the religious objections to the war of Welsh Nonconformists, some of whom became c.o.s but whose churches — aside from a few tiny millenarian sects—remained solidly pro-war. The support and leadership of the British war effort of Russell's political nemesis, David Lloyd George—a national hero in Wales and an embodiment of the historic but loosening bond between Nonconformity and Welsh Liberalism—was crucial in the latter regard. Eirug turns next to the peace politics of the Independent Labour Party (ILP) in Wales and the syndicalist wing of the South West Miners' Federation. In so doing he also presents case studies of two strongholds of anti-war sentiment: Briton Ferry and Merthyr Tydfil, both of which were visited by Russell on his July 1916 speaking tour of South Wales. The former town, a centre of tinplate production, "became a magnet for anti-war speakers from other parts of Britain" and was unflatteringly tagged "little Germany" (p. 68). The term "Merthyrism", meanwhile, was coined in *The Times* (possibly by the same febrile correspondent later taken to task by Russell in Merthyr's thriving ILP weekly: see n. 9 below) to denote a threatening conjunction of anti-war protest with industrial strife (p. 86). From a metropolitan perspective, this combustible political mix appeared unusually prevalent in this steel town in the heart of the Welsh coalfield— "cradle of the industrial revolution and the birthplace of democratic politics in Wales" (p. 78), where ILP founder Keir Hardie sat as a Labour MP from 1900 until his death in 1915.[1] Chapter 3 is devoted to the organizational work of the two main anti-conscription bodies in Wales— the No-Conscription Fellowship (NCF) and the National Council for Civil Liberties[2] (whose Welsh wing enjoyed considerable success in bridge-building to the labour movement). Finally, Eirug details the diverse experiences of the 900 or so Welsh c.o.s, the vast majority of whom (all bar about 70) accepted some form of alternative service in various civilian and military settings.

Eirug has mined the contemporary Welsh and English newspaper and periodical literature to great effect. He has also consulted many archival sources, including the records of a British intelligence and internal security apparatus increasingly perturbed, as the war dragged on, by anti-war dissent and labour militancy in Wales. It is worth noting that this last trove of official documentation would have been largely inaccessible to Jo Vellacott, whose pioneering investigation of Russell's peace work[3] is cited by Eirug, or to the editors of *Collected Papers* 13 and 14. From Eirug's helpful introductory survey of a rich secondary literature, readers will learn that he intends to modify assumptions made about the fervency of Welsh patriotism during wartime. This historiography has tended to dwell on the pro-war enthusiasm of Wales in counterpoint to its pre-existing antimilitarism, grounded in the Nonconformist tradition and more entrenched than elsewhere in Britain. Eirug also questions the depiction of pacifism in Wales as largely fragmented and ineffectual and of Welsh ILP branches whose uncompromising anti-war radicalism isolated them from and undermined the labour mainstream. (The ILP's dissenting platform was ultimately embraced by Labour, which became the majority party in Wales after the general election of 1922.) Although he is quite prepared to conclude that "resistance to war was always a minority response" (p. 227), Eirug constructs a convincing picture of a robust, coherent and coordinated anti-war movement in Wales. It was firmly rooted in tight-knit communities and gained impetus from "the jettisoning of the traditional tenets of liberalism" (p. 45)— notably the imposition of conscription and the looming threat of compulsion in industry—as well as war-weariness and a sense of political possibility fuelled by the Russian Revolution.

"High-profile" anti-war figures such as Russell (or Philip Snowden and E. D. Morel, to name but two others of many) whose voices were regularly heard at political gatherings in South Wales certainly contributed to its vigorous culture of dissent. And Russell, for one, was affected quite profoundly by the politically conscious coal miners and steelworkers he encountered as he delivered more than thirty speeches across South Wales in the first three weeks of July 1916. One such individual was Ted Williams, who seems prototypical of the militant "advanced men" of the South West Miners' Federation whose anti-war and trade union activities are discussed by Eirug.[4] Williams obtained a political education at the Central Labour College in London, then lectured for the institution in Wales before returning to mining in wartime as a checkweighman at Mardy and becoming agent for the miners' union at the same colliery after the war. In later years his politics must have softened, for when Russell next met him (in Canberra in 1950), the former Labour MP for Ogmore in Glamorgan was Britain's High Commissioner to Australia. Russell was then engaged in a lecture tour *far* less risky than that undertaken 34 years previously, when both men, as he reported from Australia to his friend Rupert Crawshay-Williams, "were on the verge of going to prison" (26 July 1950; quoted in *Papers* 26: xxix).

Russell embarked upon his journey through South Wales only two days after the appeal of his recent conviction under the Defence of the Realm Act (DORA) was dismissed. During his tour he also learned that he would be prevented from travelling to the United States to lecture at Harvard and, most painfully, that Trinity College Council had deprived him of his lectureship. More reprisals followed in the wake of his speech at the Friends' Meeting House in Cardiff on 6 July—a call for immediate peace negotiations—after some undoubtedly contentious passages were publicized further by one of his persistent hecklers, Captain W. H. Atherley Jones, a Welsh army officer. The Home Office considered most of Russell's address to be in breach of the same Defence of the Realm Regulation (No. 27) under which he had just been successfully prosecuted. But no charges were laid, lest Russell again use the courtroom as a platform to propound his anti-war views.[5] Instead he was subjected to a no less draconian sanction (but administrative as opposed to judicial) severely restricting his freedom of movement.[6] Russell intended his public speaking in South Wales to be "the first stage in a nation-wide crusade" (*Papers* 13: 420). But as he was now barred from all "prohibited areas" (which covered almost the entire coastline and, among other big cities, Glasgow—another hub of labour unrest), Russell would be prevented, as the head of MI5 minuted approvingly, from further airing "his vicious tenets amongst dockers, miners and transport workers" (quoted *ibid.*, p. lxiv). Russell thought that the War Office had acted against him from the mistaken belief that he had "tried to stir up a strike among the miners in South Wales.... Of course", he told Lady Ottoline Morrell, "I did nothing of the sort" (4 Sept. 1916; *ibid.*, p. 453). But he *was* trying to drum up support for the peace movement and had spread his defiant "stop-the-war" message all over South Wales.[7] As intimated to him by two departmental officials (see 69 in *Papers* 13), it was probably this provocation that caused him to fall foul of DORA again—and not for the last time.[8]

While Russell was thus restrained, Welsh branches of the ILP and NCF (Russell's hosts at most stops on his itinerary) were also facing mounting official scrutiny and repression (Eirug, pp. 69–71, 152–4)—as well as public anger and opprobrium (which Eirug downplays somewhat). For example, it was the trial and conviction in May 1916 of two NCF members in Cefn (near Merthyr), for distributing the "Everett Leaflet" (49 in *Papers* 13), which prompted Russell publicly to declare his authorship of this anti-conscription tract and goad the authorities into prosecuting him (see 54 *ibid.*). South Wales had been a source of governmental disquiet ever since a successful miners' strike (over wages) in July 1915. Glamorgan's zealous Chief Constable regarded his jurisdiction as a hotbed of disloyalty that needed to be deterred by the exemplary punishment of the worst culprits. As a result, a number of alleged DORA offences reported by the county constabulary were tried during the first two years of the war. Subsequently, the "more emollient approach" (p. 69) of higher civilian and military authorities usually prevailed, although the latter could always fall back on sweeping extra-judicial powers vested in them by the same emergency legislation.

As Eirug notes (pp. 72–3, 154–5), Russell was exhilarated by his generally receptive working-class audiences in South Wales. Many of his dissenting peers had already written off predominantly patriotic British labour and, like the left-wing journalist H. N. Brailsford, saw hope for the fulfilment of a progressive international agenda only "in a revolt of the saner middle-class Liberals" (quoted in *Papers* 13: xxiv). But Russell acquired a more class-based outlook on the war and a new confidence in the potential for cross-class political collaboration. The following year, after accepting an offer from the Merthyr Tydfil *Pioneer* to write a monthly column, he used his first submission to this leading organ of the ILP in Wales to issue a forthright appeal to labour in apocalyptic language that foresaw (and even welcomed) class conflict:

> [E]ither Labour or Capital must ultimately go down. There will not be enough wealth in the country for both to prosper. Either the growth of Socialism will secure for Labour a more adequate share of the national wealth, or else Capital, backed by the State and the Army, will succeed in reducing Labour to a servile condition, in which wages will be only just sufficient to support life. This was the condition of Labour at the end of the Napoleonic Wars, and if our present masters have their way, it will be its condition again when this war ends.[9]

Russell felt that he was largely preaching to the converted in the Welsh mining and steel communities he visited (p. 155), but he clearly experienced a less friendly reception when he publicly attacked British war policy in Cardiff (see above). Four months later (on 11 November 1916) a much more violent display of organized rowdyism (combined with police inaction) resulted in the break-up of an anti-conscription meeting at Cory Hall in the Welsh capital. In his detailed account of this brazen challenge to public order and of its polarized social and political backdrop, Brock Millman emphasizes the "splintered working-class reaction to the war"—not only in Wales, where the politics of class and nation, previously complementary, were now at odds, but more widely across Britain.[10] Eirug focuses instead (pp. 131-2) on the impressive level of dissenting participation achieved when the disrupted event was finally staged in the largest indoor venue in Merthyr a few weeks after the Cardiff fiasco.

Popular patriotic hostility (and government surveillance and police harassment) did not suppress the surge of political optimism that spread through the Welsh peace movement after Russian Tsarism was overthrown in March 1917. Russell too participated in the ensuing "Summer of Hope" (see *Papers* 14, Pt. v). Buoyed by the largely non-violent end of Tsarist rule and its replacement by a provisional government determined to leave the war, Russell was persuaded (albeit only briefly) that all warring states could forge a similar synthesis of pacifism and revolution. On 3 June he addressed the storied Leeds Convention (40 *ibid*.), held in solidarity with the Russian Revolution and whose delegates included a militant socialist contingent from Wales. To build on the radical and anti-war momentum generated at Leeds, follow-up meetings were arranged in several of the Welsh towns where Russell had spoken in July 1916, while district councils of workers' and soldiers' deputies were to be set up throughout Britain. Indeed, Russell was elected to the London body, only to witness the riot which prevented its inaugural gathering from taking place in Southgate's socialist and pacifist Brotherhood Church (see 61 ibid.). The following day (29 July) a similarly violent fate befell the founding conference of the Welsh district council in Swansea, and any prospect of British Soviets—always chimerical perhaps, as Eirug implies (pp. 101–2)—quickly dissipated.

The chapters on the anti-conscription struggle and the travails of Welsh c.o.s examine at a granular level issues that perplexed Russell as acting chair of the NCF's National Committee throughout 1917: tensions between "absolutists" and "alternativists", harsh prison terms and conditions inflicted on the former, unrest fomented by the latter in quasi-penal Home Office work camps, and the role of local tribunals and other administrative bodies in interpreting and implementing the "conscience clause" of the Military Service Acts.[11] The most fundamental dilemma for the c.o. movement, abundantly clear in the Welsh context so meticulously reconstructed by Eirug, was preserving the common purpose of such a theologically and politically disparate group. For "quietist" religious c.o.s, refusing to enlist was often an expression only of an individual peace witness. "Whilst all who opposed the war opposed the extension of conscription", as Eirug puts it, "not all who opposed conscription opposed the war" (p. 160). But the politically committed resistance to conscription strove (with only limited success) to hitch this campaign to a broader peace effort. Russell and others wanted to end not only the suffering and hardship of the c.o.s but the war itself. For readers with a particular interest in these and other pivotal episodes in Russell's eventful political life during World War I, or those simply curious about modern Welsh history, *The Opposition to the Great War in Wales* has much to offer.

*Andrew Bone*

**Works cited**
Bone, Andrew G. "Russell and the Other DORA, 1916-18". *Russell* 38 (2018):108-9.
Millman, Brock. *Managing Domestic Dissent in First World War Britain*. London: Frank Cass, 2000
Russell, Bertrand. *Two Years' Hard Labour for Refusing to Disobey the Dictates of Conscience*. [The Everett Leaflet.] London and Manchester: National Labour P., 1916. 49 in *Papers* 13.
—. "The Cardiff Speech" (1916); 63 in *Papers* 13.
—. "The Times on Revolution", The Pioneer, no. 342 (6 Oct. 1917): I; 74 in *Papers* 14.
—. *The Brixton Letters*. Ed. Kenneth Blackwell, Andrew G. Bone, Nicholas Griffin, Sheila Turcon. 2018. CLBR, russell-letters.mcmaster.ca.
—. *Auto*. I.
—. *Papers* 13, 14, 26.
Vellacott, Jo. *Conscientious Objection: Bertrand Russell and the Pacifists in the First World War*. Nottingham: Spokesman, 2015; 1st ed., 1980.

**Notes**
1. In the ensuing by-election, however, the ideological fault-line cut by the war through working-class Wales was glaringly revealed. The ILP dissenter who was Labour's official candidate was defeated by the fervently pro-war miners' agent (i.e. trade union official) who ran against him (also for the ILP) after losing a bitterly fought nomination contest. The victorious C. B. Stanton, MP, later led the Cardiff mob that attacked a public protest of conscription in the so-called "Battle of Cory Hall" (see Millman, *Managing Domestic Dissent in First World War Britain*, Ch. 6).
2. Formed as the National Council against Conscription shortly after the *first* Military Service Act became law in January 1916.
3. I.e. *Conscientious Objection: Bertrand Russell and the Pacifists in the First World War* (2015; 1st ed., 1980).
4. He is not positively identified by Eirug but may well be the "Ted Williams" who is mentioned in passing (p. 149) in connection with the Pontypridd branch of the NCF.
5. See the reviewer's "Russell and the Other DORA, 1916-18" (2018), pp. 108-9.
6. By a War Office order dated 1 September 1916 and issued under Defence of the Realm Regulation No. 14.
7. A Home Office transcription of Russell's Cardiff address (63 in Papers 13), prepared from shorthand taken by a reporter for the city's *Western Mail*, is regrettably the "only extant account of a complete speech on his Welsh tour" (ibid., p. 420). The *Pioneer's* reports of meetings addressed by Russell in Tai-bach, Merthyr and Abercanaid (8 July 1916, p. 2 and 15 July 1916, p. 4) indicate that he spoke in the same political vein at these events as in Cardiff.
8. See *The Brixton Letters*.
9. *"The Times* on Revolution" (1917), p. I (74 in Papers 14).
10. *Managing Domestic Dissent*, pp. 138-47 (quotation at 139).
11. See *Papers* 14, Pts. II and IV especially.

*We are grateful to Andy Bone and to* russell: The Journal of Bertrand Russell Studies *for permission to reprint this review.*